ELECTRICITY MATTERS

Englisch für elektrotechnische Berufe

von
Michael Benford
Wolfgang Towara

Cornelsen

Electricity Matters New Edition wurde geplant und entwickelt von der Redaktion Moderne Fremdsprachen des Cornelsen Verlags, Berlin.

Verfasser:	Michael Benford, Bochum Wolfgang Towara, Würzburg
unter Mitarbeit von:	Wolf-Rainer Windisch, Vörstetten
Berater:	Roland Kiesel, Schwebheim Wolf-Rainer Windisch, Vörstetten
Projektleitung:	Jim Austin
Verlagsredaktion:	Andreas Goebel
Außenredaktion:	Steve Williams, Corio Bay Media, Melbourne
Redaktionelle Mitarbeit:	Laura Spratling; Fritz Preuß (Wörterverzeichnisse)
Bildredaktion:	Gertha Maly
Layout und technische Umsetzung:	Oxford Designers & Illustrators
Umschlaggestaltung:	Katrin Nehm
Titelfoto:	Corbis/RF
Illustrationen:	Oxford Designers & Illustrators

Erhältlich sind auch:
Audio-CD
Handreichungen für den Unterricht

www.cornelsen.de

Die Links zu externen Webseiten Dritter, die in diesem Lehrwerk angegeben sind, wurden vor Drucklegung sorgfältig auf ihre Aktualität geprüft. Der Verlag übernimmt keine Gewähr für die Aktualität und den Inhalt dieser Seiten oder solcher, die mit ihnen verlinkt sind.

Dieses Werk berücksichtigt die Regeln der reformierten Rechtschreibung und Zeichensetzung.

1. Auflage, 3. Druck 2010

Alle Drucke dieser Auflage sind inhaltlich unverändert und können im Unterricht nebeneinander verwendet werden.

Druck: orthdruk, Bialystok, Polen

ISBN 978-3-464-20795-6

Inhalt gedruckt auf säurefreiem Papier aus nachhaltiger Forstwirtschaft.

VORWORT

Electricity Matters New Edition ist die völlige Neubearbeitung des bewährten Englisch-Lehrwerks für Lernende elektrotechnischer Berufe, die am Berufsschulunterricht oder an betrieblichen Fortbildungsmaßnahmen teilnehmen. Das Werk basiert auf den Lehrplänen der Bundesländer für Berufsschulen und bereitet auf die KMK-Zertifikatsprüfung Stufe II vor (Niveau B1 des Gemeinsamen Europäischen Referenzrahmens).

Electricity Matters New Edition richtet sich an Lernende, deren Englischkenntnisse dem Niveau eines mittleren Bildungsabschlusses entsprechen. Darüber hinaus enthält jede Unit mit den *Extra materials* auch anspruchsvolleren Lernstoff für Lernende mit fortgeschrittenen Englischkenntnissen. Ein Grundwortschatz von ca. 650 Wörtern wird vorausgesetzt. Dieser ist in einem gesonderten Vokabelverzeichnis aufgeführt.

Electricity Matters New Edition besteht aus zwölf Units, von denen jede eine in sich abgeschlossene Handlung enthält. Es ist möglich, einzelne Units auszulassen, jedoch wird empfohlen, wann immer Zeit und Lehrpläne es ermöglichen, die gegebene Reihenfolge einzuhalten, um einen kontinuierlichen Aufbau des Fachvokabulars zu gewährleisten.

In berufsbezogenen, praxisnahen Situationen werden die Lernenden mit allen wesentlichen Aspekten des Englischen vertraut gemacht, die ihnen in ihrem beruflichen Alltag begegnen können. Ob es um Kontakte zu Kunden und Geschäftspartnern, den Umgang mit technischen Dokumenten, das Schreiben von E-Mails, die Kommunikation über Bauteile, Messungen oder Motoren geht – **Electricity Matters New Edition** bietet in jedem Fall das passende sprachliche Rüstzeug.

Alle Units zeichnen sich durch kleine Lernschritte aus, so dass auch Lernende, die nur eine Wochenstunde Englischunterricht haben, ihre Kenntnisse und Fähigkeiten sorgfältig und zielstrebig aufbauen können.

Electricity Matters New Edition bietet die Möglichkeit eines handlungsorientierten und praxisbezogenen Unterrichts. Die aus dem Berufsleben gegriffenen, häufig authentischen Materialien vermitteln den Lernenden praxisnah das wichtige Fachvokabular. Die Lernenden schulen aktiv ihre Sprechfertigkeit in Rollenspielen und in Partnerarbeit, wobei neben der beruflichen auch die soziale Komponente des Arbeitsalltags einbezogen wird.

Grammatik wird in jeder Unit punktuell wiederholt, wobei die Regeln auf das Wesentliche begrenzt sind und insbesondere der Festigung von Grundstrukturen dienen.

Die Rubrik *Learning* bietet in jeder Unit Lerntechniken, die gezielt zeigen, wie man sich den neuen Wortschatz aneignet oder beim Verstehen und Schreiben sicherer werden kann. *Culture*-Kästen weisen auf interkulturelle Besonderheiten hin oder vermitteln landeskundliches Hintergrundwissen.

Electricity Matters New Edition vermittelt die englischsprachigen Grundpfeiler, auf denen sicheres Auftreten und korrektes Handeln im Berufsalltag basieren. Wenn die Lernenden sich dabei mit Vergnügen der englischen Sprache bedienen, hat dieses Werk sein Ziel erreicht.

Table of contents

UNIT 1
My company

1 WARM UP

Work with a partner. Talk about your job. Include the following information.

1 Describe the type of work your company does.
It sells … / wires … / does … work.
2 Does it have private, commercial or industrial customers?
3 What sort of work does it specialize in?
4 How many employees does your company have?
5 Where is it located?
It's located in … / near … / 50 km from … / between … and …
6 How many trainees / qualified electricians / master electricians / electronics technicians does it have?
7 Describe what you do there in a few sentences.

2 A NEW CONTRACT FOR FISCHER ELEKTRONIK

Eddie Waterhouse is a British businessman. His company, Waterhouse Paints, plans to open a paint distribution centre in Essen. He receives an email from Fischer Elektronik.

Read the email and say if statements 1–6 are true or false. Correct the false statements.

1 Eddie works for Fischer Elektronik.
2 Eddie plans to open a paint distribution centre in Essen.
3 Franziska Drechsler works for Fischer Elektronik.
4 Fischer Elektronik is a new company.
5 Fischer Elektronik has a good name.
6 Jochen wants to meet Eddie in England.

Von: jochen.kowalski@fischerelektronik.de
Cc:
Betreff: Electrical installation work

Dear Mr Waterhouse

Frau Franziska Drechsler, your architect, tells me that you are looking for a contractor to do the electrical installation work for a paint distribution centre in Essen.

My company has an excellent reputation both at home and abroad for electrical work of the highest standards, performed on schedule and to the satisfaction of the customer.

As we have 20 years' experience in this field I feel sure that we are the business partner you are looking for and would therefore be delighted to meet you and discuss further details with you in person when you come to Germany.

Please phone me on +49 201 9455347 or send me an email to arrange a meeting.

I look forward to hearing from you soon.

Yours sincerely

Jochen Kowalski
Managing Director

LEARNING

Wie kann ich mein Englisch verbessern?

Englisch kann man auf viele unterschiedliche Weisen lernen. Hier sind ein paar Tipps – aber bestimmt fallen Ihnen noch mehr Möglichkeiten ein!

- Filme in Originalfassung sind mit Untertitel leichter zu verstehen – und vielleicht klappt es beim zweiten Mal schon ohne Untertitel!
- Ein Magazin für Ihr Hobby gibt es ganz bestimmt auch in englischer Sprache.
- Warum nicht mal eine SMS oder E-Mail auf Englisch schreiben?
- CDs haben meist Booklets mit Texten – wer mitliest, versteht mehr.

3 A MEETING WITH THE ARCHITECT

2 Eddie Waterhouse and the architect, Franziska Drechsler, meet Jochen at his office at Fischer Elektronik.

Read or listen to the dialogue. What does Eddie want from Fischer Elektronik? Make notes in German.

FRANZISKA (*To Eddie*) Let me introduce you to Mr Fischer.
EDDIE Pleased to meet you.
JOCHEN Pleased to meet you, Mr Waterhouse. (*They shake hands.*)
EDDIE Please call me Eddie.
JOCHEN And please call me Jochen.
FRANZISKA Eddie's company wants to set up a paint distribution centre here and he'd like to discuss some of the details.
EDDIE Yes, that's right. We're talking about 1,000 m^2 of storage space and 300 m^2 of office space. We'll need bright lighting in the warehouse area because it helps to reduce packing errors. Does your company normally do that sort of work?
JOCHEN Yes, we do. We regularly get contracts for that sort of thing. Anyway, we'll certainly need explosion-proof fittings.
EDDIE Is that a regulation?
JOCHEN Yes, it is. All the switches, plugs, lighting elements and so on have to be airtight so a spark doesn't set off an explosion.
EDDIE Oh, I see – but it sounds expensive!
JOCHEN I'm afraid we have no choice in this case. Look, I've got an idea. We're doing the installation work for a petrol station near here at the moment and we're using the same fittings that you'll need in your distribution centre, so would you like to visit the site and see what we're installing?
EDDIE Yes, that sounds like a good idea.

CULTURE

Short answers

Im Englischen gilt es als unhöflich, nur mit *Yes* oder *No* zu antworten. Stattdessen gibt man Kurzantworten, die in der Regel aus drei Wörtern bestehen:
Yes, we do. / No, we don't. / Yes, it is.

Häufig dienen Standardformulierungen dazu, Ja/Nein-Antworten höflicher klingen zu lassen:
Yes, that sounds like a good idea. / Yes, that's right. / No, I'm afraid we …

4 UNDERSTANDING THE EMAIL AND DIALOGUE

Complete the following statements about Jochen's email and his conversation with Eddie.

1 Eddie is looking for …
2 Eddie's architect told Jochen …
3 Fischer Elektronik does …
4 The paint distribution centre …
5 In the warehouse, …
6 Packing errors often happen because …
7 The fittings …
8 Jochen thinks it would be helpful …

5 WORKING WITH WORDS

Use the email and the dialogue to find words or phrases which match the definitions or fill the gaps.

1 A company which accepts contracts to do work is called a … .
2 (good) name
3 at the time expected
4 The … is the head of the company.
5 A place where goods are stored and sent out to customers is a … .
6 a mistake
7 Switches, plugs and sockets are all electrical … .
8 to cause (e.g. an explosion)

GRAMMAR

Talking about the present: simple present and present progressive

1 We**'re talking** about 1,000 m^2 of storage space.
2 We**'re doing** the installation work at the moment.
3 We regularly **get** contracts.
4 **Does** your company normally **do** that sort of work?
5 … so a spark **doesn't set off** an explosion.

- Das *Present progressive* wird gebraucht, wenn man gerade ablaufende oder noch nicht abgeschlossene Handlungen beschreibt. Es steht oft in Verbindung mit Signalwörtern wie *at the moment, now, (right) now* oder *just.* (1, 2)
- Das *Simple present* wird für wiederholte, oft regelmäßige Handlungen in der Gegenwart verwendet. Es steht oft in Verbindung mit Signalwörtern wie *regularly, normally, often, sometimes* oder *always.* (3–5)
- Manche Verben, die keine Handlung, sondern einen Zustand ausdrücken (z.B. *need, like, want, hate, love*), stehen immer im *Simple present.*
- In der 3. Person (*he, she, it, the man* usw.) wird ein *-(e)s* angefügt. Frage und Verneinung werden mit einer Form von *to do* (*do/does, don't/doesn't* + Grundform) gebildet. (4, 5)

6 PRACTICE

Complete the text with the correct verb forms.

Eddie Waterhouse's company … (sell)[1] paint over Europe and now in Germany. Eddie … (do)[2] his best to find an electrical contractor for the new distribution centre. He regularly … (fly)[3] long distances and … (not spend)[4] much time at home. At the moment he … (talk)[5] on his mobile to Patsy, his secretary.

PATSY Can you talk now, Mr Waterhouse or … (you / drive)[6]?
EDDIE It's all right. I … (just / walk)[7] into my hotel.
PATSY Good. Now Tim Watson in the sales division … (need)[8] to speak to you. He … (work)[9] on his report at the moment and he's got a problem.
EDDIE I see. … (he / want)[10] me to call him?

7 FISCHER ELEKTRONIK'S WEBSITE

a **Fischer Elektronik has translated its website into English but the headings and texts have got mixed up. Match the texts with the headings.**

File Edit View Go Bookmarks Tools Help

http://www.fischerelektronik.de/

1 Electrical Installation	More than 80% of the information we receive comes through our eyes. Whether it's a restaurant or a railway station we have the solution.
2 Network Technology	Reduce your energy costs by using energy from the sun! A photovoltaic system could be the solution for you.
3 Solar Technology	Our 24-hour hotline puts you in touch with our mobile team of qualified technicians who will solve your problem without delay.
4 Communications Technology	Our teams of experts will use state-of-the-art technology and know-how to plan, install, wire, rewire and modernize your electrical systems.
5 Lighting Technology	Fibre optics or copper cables? We'll plan, install and service the computer system your company needs.
6 After-sales Service	No more missed calls! Our reliable systems come at a price you can afford. We'll also install digital TV (DVB) and radio (DAB) for you.

b **Now match the pictures with the headings (1–6).**

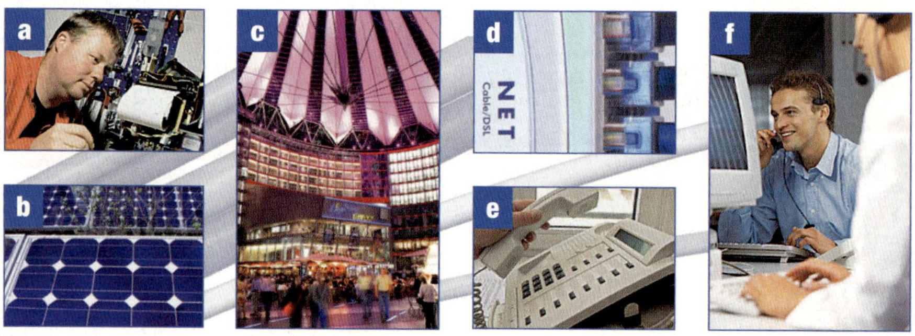

8 USING THE RIGHT WORD

Complete the sentences with the words from the box. They are all from the website above.

service ■ mobile ■ photovoltaic ■ reduce ■ reliable ■ rewire ■ solution ■ state-of-the-art

1 Fischer Elektronik installs solar energy systems using … cells.
2 Customers like them because they … the amount of money spent on electricity.
3 They are the most modern systems on the market. They are … .
4 Fischer Elektronik's technicians always come when they are called. They're … .
5 They always find the right … to the customer's problem.
6 Each technician has his own van so he is … and can travel anywhere.
7 Electrical systems don't break down often if you … them regularly.
8 Sometimes the cables in a building are so old that you have to … the whole building.

9 USING AN ORGANIZATION CHART

Study the organization chart of Fischer Elektronik and then correct the statements below.

1. Fischer Elektronik employs 50 people.
2. It has four administrative and five technical sections.
3. The electrical installation section is the smallest section with 20 employees.
4. The solar technology section employs most people.
5. There are five directors under the managing director.
6. The warehouseman is in charge of after-sales service.
7. The directors each have a secretary.
8. The technicians in the communications technology section put in household wiring and repair electric motors.

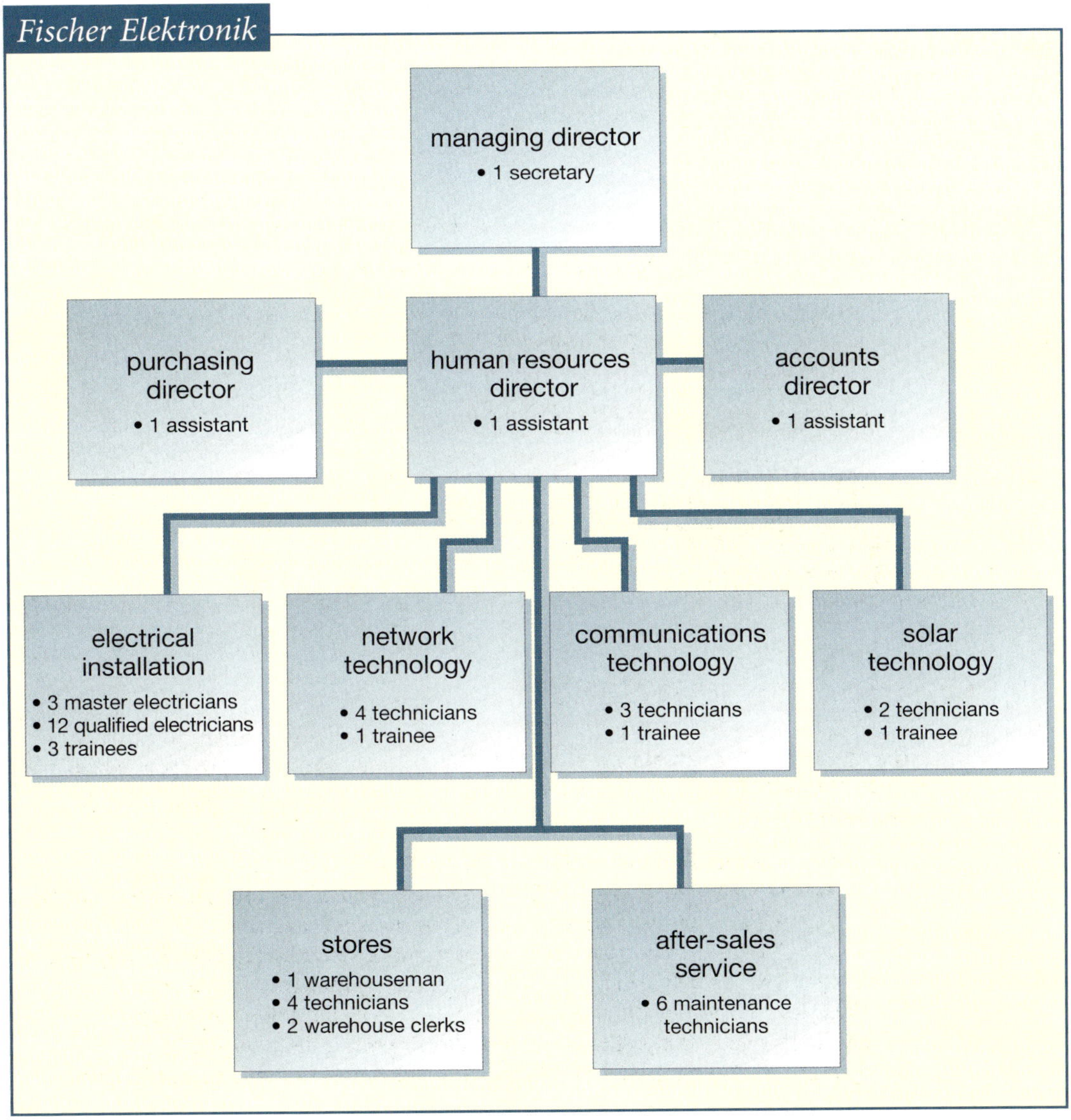

10 TALKING ABOUT YOUR JOB

a Read what these five employees of Fischer Elektronik say about their jobs. What are their jobs called and which departments do they work in?

Sometimes it's monotonous – you know, just putting in cables and wiring light switches and so on – but later, when I'm qualified, I'll be telling people what to do!

My boss is very nice but I wouldn't like his job because he has to hire and fire people. It can be very stressful.

I've always loved connecting up computers – I remember those all-night LAN parties in the cellar when I was 14 – and now I go from firm to firm and get money for doing the same thing. I love it!

You drive around a lot and meet people in their homes or at work but you have to know what you're doing, because the customer wants to know what's wrong immediately and how much it'll cost to repair it.

My last job was a big DIY store where all the assistants had to have cordless phones and be on call all the time. Another interesting job was installing the bleepers for hospital doctors.

b Now invent descriptions for two other employees at Fischer Elektronik. Can your partner guess what their jobs are called and where they work?

11 TALKING ABOUT YOUR COMPANY

a With a partner, draw up an organization chart of your own company.

b Work in groups of 2–4. Use the phrases below to write a description of your company and what you do there. Add other important points. Give a presentation to the class.

I	work for a	small medium-sized large		firm company	based in	… (town, city, e.g. Munich).
We	work with have	private business all sorts of		customers	in the local area. all over Germany. all over Europe.	
My company	employs	… (number, e.g. 25)		people	including … (number)	engineers. qualified electricians. trainees. office staff.
We	wire rewire	houses offices shops factories	and	install maintain repair replace	electric motors electricity generators electrical equipment	in … (e.g. private homes, shops, hospitals, etc.).

12 YOUR COMPANY ON THE WEB

Invent a simple website for your company with headings about its main activities. Write a short text as a link under each heading. Use Fischer Elektronik's website (see exercise 7) as your model.

13 A GUIDED TOUR OF FISCHER ELEKTRONIK

3

Listen to the dialogue between Jochen Fischer and Eddie Waterhouse. Then match a–l with the numbers on the diagram. (Start with number 1.)

- a Maintenance and delivery vehicles car park
- b Administration building
- c Canteen
- d Warehouse
- e Reception
- f Employees' and visitors' car park
- g Communications technology
- h Lighting technology
- i Electrical installation
- j After-sales department
- k Network technology
- l Solar technology

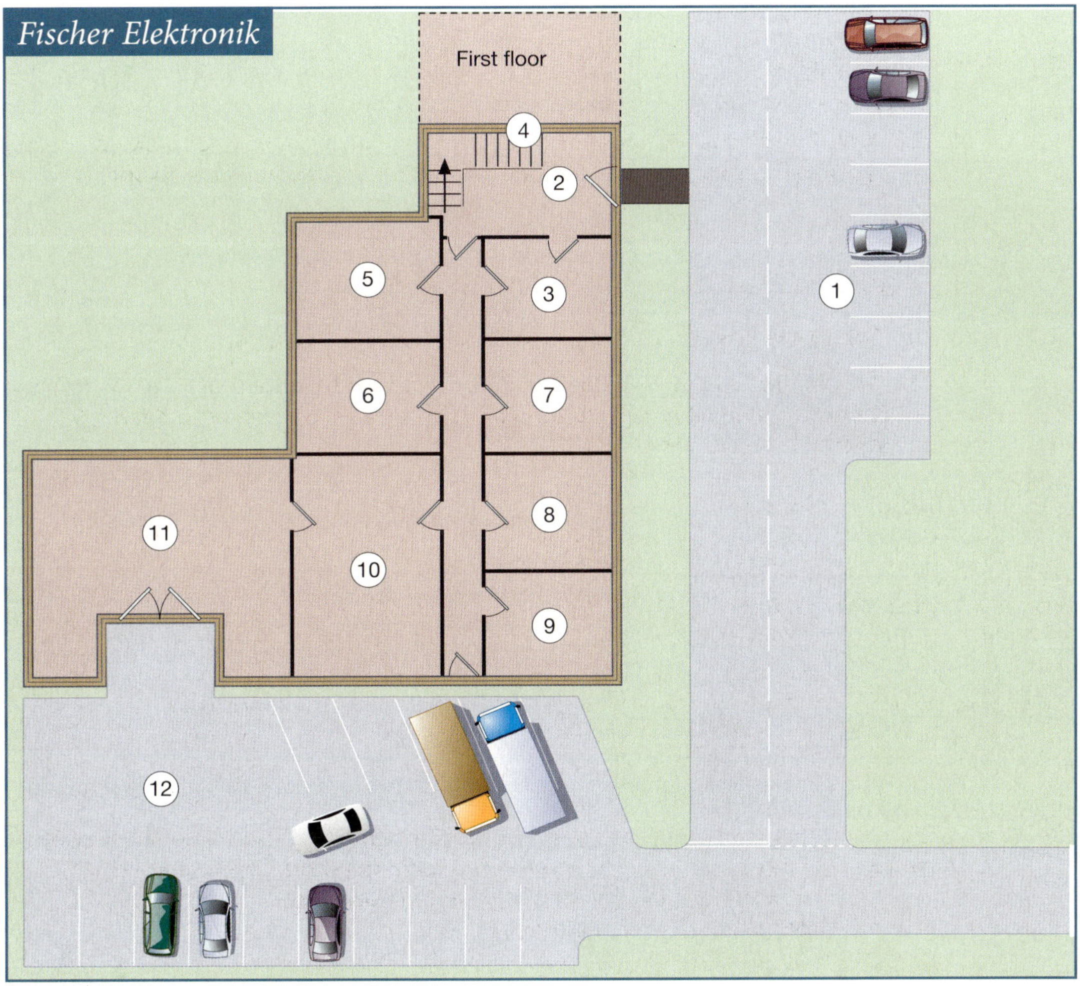

14 TRUE OR FALSE?

3

Now listen to the dialogue again and say if the following statements are true or false. Correct the false statements.

1 Jochen drives Eddie to Fischer Elektronik's site.
2 Eddie will go back to the hotel after this visit.
3 The canteen is in the administration building.
4 Two technicians work in the solar technology section.
5 Solar technology is less popular nowadays.
6 The two employees they see in the electrical installation section are unqualified.
7 Fischer Elektronik installs more TV and radio aerials than phone systems.
8 Fischer Elektronik only has a small stock of electrical supplies.

15 SHOWING A VISITOR AROUND THE COMPANY

An English-speaking visitor is in reception at Fischer Elektronik and wants to know the way to:

1 the electrical installation section
2 the after-sales section
3 the warehouse
4 the maintenance and delivery vehicles car park
5 the canteen

Act out the situation with a partner using the expressions in the box.

Begin like this:

A: *Excuse me. Could you tell me the way to the electrical installation section, please?*

B: *Yes, certainly. Go through reception and turn left into the main corridor. It's the …*

LANGUAGE

Giving directions

Go to/through/past …
Turn left/right into …
The … is on your left/right.
next door to
opposite
outside
in front of
at the end of
on the other side of
the next/first/second door on your left/right

16 NOW IT'S YOUR TURN

With a partner use a plan of your own company to help a visitor to find his or her way around.

Apply Online for a Job in Hanover TODAY!

Company: Avionics Engineering Inc.
Location: Hanover, Pennsylvania
Rate: $17.65 per hour

Electronics Technician Job Description:

Avionics Engineering Inc requires two maintenance electricians. Candidates should be able to fabricate, install and maintain all electrical systems and panels. This includes repairs, installation, and testing electrical circuits, working with high voltage lines. The right candidate should have experience in the electrical field, preferably in a maintenance field.

One shift is Friday–Sunday 6 a.m.–6 p.m., the other is Friday–Sunday 6 p.m.–6 a.m. Anything over 8 hours each day is considered overtime. Please submit a résumé for consideration. Only qualified candidates will be invited to interview.

Required Skills for Electronics Technician Job:

- ELECTRONIC REPAIR/MAINTENANCE
- ELECTRICAL MAINTENANCE
- ELECTRICAL

About our company:

Our terms of employment include comprehensive medical, dental and optical benefits, an optional, government-sponsored pension plan, a competitive wage and solid career advancement opportunities.

Click here to apply

1 QUESTIONS

1 **What type of work does an avionics engineering company do?**
2 **How does the hourly wage offered here compare with German pay rates?**
3 **Look at the skills needed to do this job and then say if the training course you are doing now covers them. What skills are not covered?**
4 **Explain the following in your own words:**
   a medical benefits
   b dental benefits
   c optical benefits
   d pension plan

2 DISCUSSION

1 **What disadvantages are there to this job?**
2 **Would you be interested in this job? Explain why (not).**
3 **Describe your ideal job in a different country.**

UNIT 2
My job and workplace

1 WARM UP

Work with a partner.

a Match the correct job title to each photo.

installation electrician
panel building electrician
repair and maintenance electrician
street lighting electrician

1
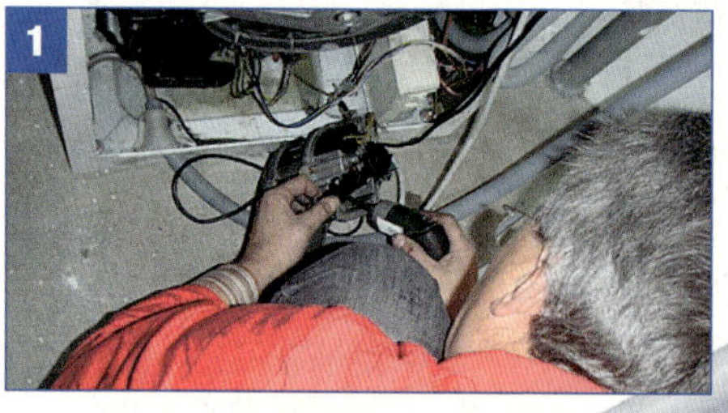

2

3
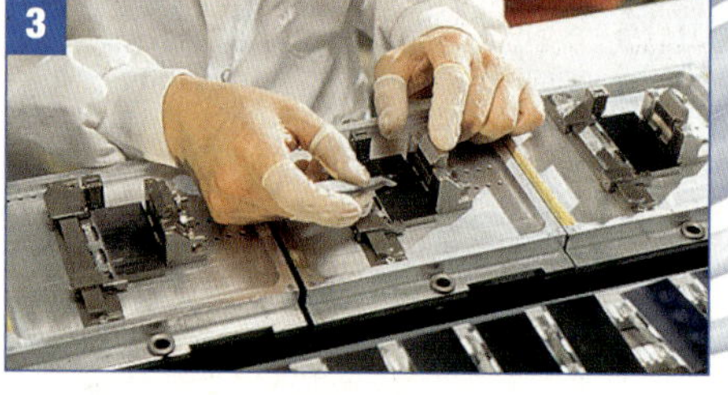

4

b What does each electrician's job involve? Make sentences using the table. You can add your own ideas.

assemble			domestic households
install	equipment	in	commercial buildings
inspect	wiring systems		industrial plants
maintain			the street
repair			a factory

2 TALKING ABOUT MY JOB

4

Four British electricians from Newcastle upon Tyne in the northwest of England are talking about their jobs.

Read or listen to the descriptions, then decide what each person's job title is. (The job titles are all in exercise 1 above.)

I'm **Jamie Wilson**. My job is to inspect, test and install wiring systems in domestic households. This involves stripping out the old systems and replacing them with new wiring. In new houses I follow architects' drawings which show where to install the electrical equipment. I also have to bend and cut conduit and fix equipment to walls.
I enjoy my job because I'm not in an office or a workshop and there's lots of variety in my work. Last week I did a job in a block of flats and this week I'm working in a 400-year-old house!
I travel around a lot depending on where the job is. My standard working week is 40 hours, but there's lots of overtime.

I'm **Ben Clark**. I follow wiring plans to assemble electronic control panels which use programmable logic controllers (PLCs). The control panels are used in commercial buildings to control the heating, air-conditioning and ventilation systems. I work in a workshop. At college I specialized in panel building: it's interesting work because the panels are very complex. I work a 37-hour week, Monday to Friday. That's good because I like regular hours.

I'm **Gill Strawford** and I'm responsible for installing, maintaining and testing electrical equipment in a big industrial plant. I repair electrical and electronic equipment when breakdowns occur. I replace items such as circuit breakers, fuses, switches and so on. I also periodically inspect all equipment to make sure it's operating properly, and locate and correct problems before breakdowns occur. One of my main tasks is to troubleshoot equipment like motors to find out why the unit isn't functioning properly. I work shifts because the industrial plant operates 24 hours a day, so there always has to be an electrician available to deal with any problems. Last week I worked nights, for example.

My name's **Neil Gillespie**. My job's to test circuits and diagnose and repair faults in street lighting and traffic management equipment like traffic lights and motorway message signs. I normally work from mobile high access platforms. I mostly work in the open air, of course, and at heights, but I don't mind that. In fact, it's what I like about my job. Sometimes working conditions are a bit inconvenient, especially when it's raining or cold. But no matter what the conditions are like, the job has got to be done. I've got fixed working hours, but at other times I'm 'on call' – in an emergency I have to go to the worksite at any time of the day or night. Last night I had to go out after midnight because a set of traffic lights malfunctioned. I didn't get home until 10 a.m.

3 UNDERSTANDING THE TEXT

Read the text again, then answer the questions. There may be more than one answer to each question.

Who …

1 carries out routine maintenance?
2 works on site?
3 inspects wiring systems?
4 works different hours on different days?
5 always works the same hours?
6 sometimes has to go to work at strange times?
7 works in a workshop?
8 assembles electrical equipment?

4 WORKING WITH WORDS

Complete the sentences with the English equivalents of the words in brackets. The words are all in the text on page 17.

1 This … (*Schalttafel*) controls the heating for the whole building. (Ben)
2 Can you come in to work, please? There's an … (*Notfall*) and we need you immediately. (Neil)
3 We need you to … (*Störungen suchen*) the production line: we don't know what the problem is. (Gill)
4 Simon works at … (*Industrieanlagen*) like power stations and oil refineries. (Gill)
5 Are you available to do some … (*Überstunden*) this week? (Jamie)
6 When there's a … (*Panne*) at the plant, it's my job to fix it. (Gill)
7 I don't work … (*an der freien Luft*) – I work indoors. (Neil)
8 … (*Abwechslung*) is important in my work – I don't like to do the same thing every day. (Jamie)

5 JOB COMPARISON

Pick the job profile (page 17) which is most similar to your job. What is the same? What is different? Make two lists.

6 MY JOB PROFILE

Use your notes and the Language box below to write a profile of your own job.

LANGUAGE

Describing jobs

I work in a workshop/factory/office.
I work on site / in the open air / at heights.
I install electrical equipment/devices.
I test/replace components.
I inspect/troubleshoot electrical systems.
I maintain/repair machines.
I diagnose faults.
I advise customers.
I follow technical drawings / instructions / wiring plans.
I work fixed hours / a 37-hour week / shifts / nights.
I can do lots of overtime.
I'm on call.

CULTURE

Urlaub und Arbeitsstunden im Vergleich

In Deutschland haben es Arbeitnehmer mit 28–30 bezahlten Urlaubstagen im Jahr vergleichsweise gut. In Großbritannien sind es ca. 20, in anderen EU-Ländern sind es zwischen 25 und 30. In den USA hingegen fangen Arbeitnehmer mit ca. 10 Urlaubstagen an, und japanische Arbeitnehmer bekommen ganze neun Tage bezahlten Urlaub.

Vergleicht man die durchschnittlichen Arbeitsstunden pro Jahr, so treten die Unterschiede noch deutlicher hervor. In japanischen Betrieben leistet der Durchschnittsarbeitnehmer 1.990 Stunden jährlich, während man in den USA 1.904 Stunden, in Großbritannien 1.774 und in Deutschland nur 1.573 Stunden pro Jahr arbeitet. In den USA und in Japan sind 50-Stunden-Wochen und mehr keine Seltenheit: Für knapp ein Drittel der männlichen Arbeitnehmer gehören sie sogar zur Norm.

Der globale Konkurrenzkampf setzt europäische – darunter auch deutsche – Unternehmen zunehmend unter Druck, die Arbeitswoche zu verlängern, ohne die Löhne entsprechend zu erhöhen.

7 OVER TO YOU

Class discussion. Form two equal groups:

Group A: You are in favour of long holidays and short working hours.
Group B: You are in favour of shorter holidays and longer working hours to protect your jobs.

1 Prepare in your group: brainstorm arguments to support your point of view.
2 Now carry out the discussion. Which group can produce the best arguments? Use expressions from the Language box.

LANGUAGE

Discussions

Beginning a discussion
Yesterday I read a web page about …
Have you ever thought about … ?
Did you know that … ?

Agreeing
I agree with that.
That's what I think too.
You're absolutely right.

Disagreeing
I'm afraid I don't agree.
I think you're wrong about that.
Sorry, but that isn't correct.

Giving an opinion
In my opinion …
Can I just point out that … ?
Well, if you ask me …

Doubting
I'm not so sure.
That may be true, but …
I have a few doubts about that.

Ending a discussion
We've heard some interesting points, so let's stop here.
We'll just have to agree to disagree.

LEARNING

Flüssiges Sprechen trainieren: Shadow reading

Nehmen Sie Nachrichten in englischer Sprache auf und hören Sie sich den Text an. Nun wird der Text mehrere Male wiederholt, jedoch immer leiser. Versuchen Sie bei den weiteren Wiederholungen mitzusprechen, indem Sie sich allmählich vom Text lösen, bis Sie ihn auswendig sprechen können. (Hörtexte aus diesem Buch sind eine andere gute Quelle für diese Art Übung.)

8 AT MY WORKBENCH

Match the names of the tools and equipment with the numbers in the picture.

Allen keys ■ water pump pliers
power drill ■ folding rule
hacksaw ■ hammer ■ multimeter
Phillips screwdriver ■ round nose pliers
flat-bladed screwdriver
side-cutting pliers
soldering iron ■ spanner
needle nose pliers ■ vernier calliper
wire strippers

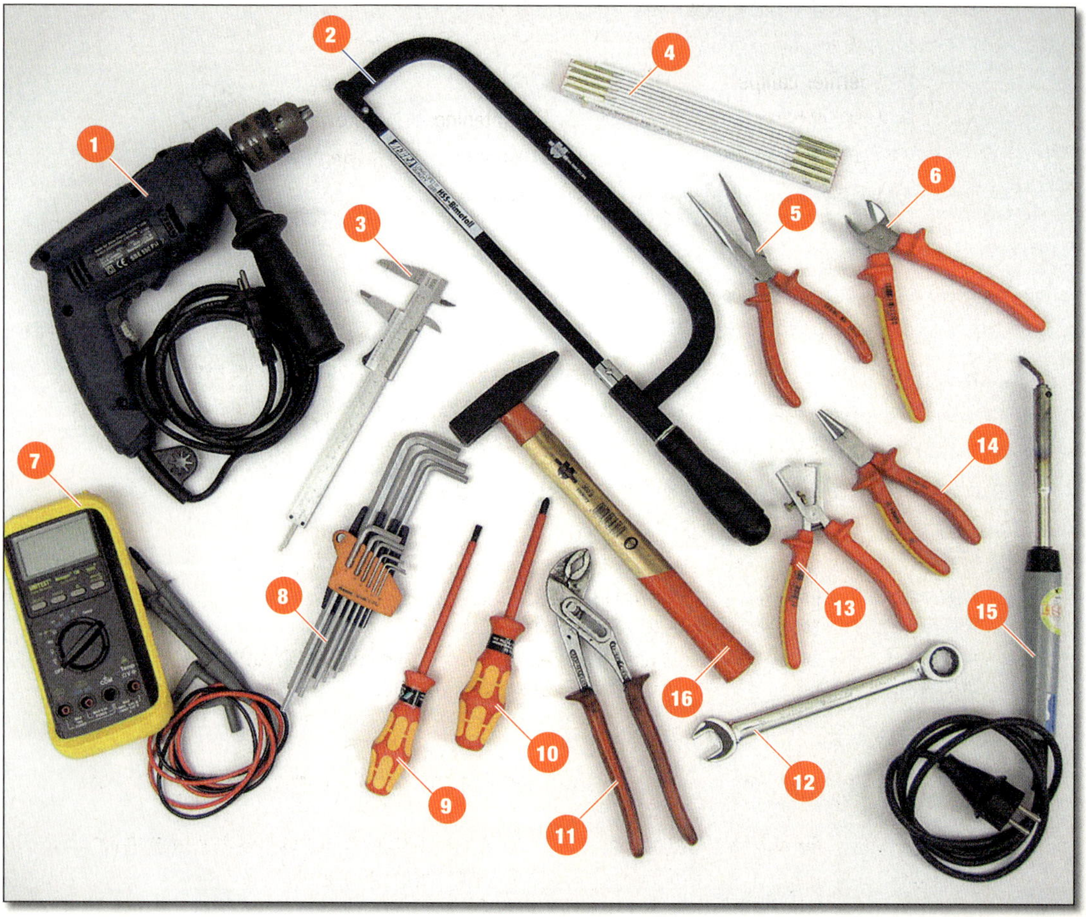

9 TOOL TALK

Discuss in a small group:

- **Which of the tools above do you use and how often do you use them?**
- **Which other tools and equipment do you use?**

10 THE RIGHT TOOL FOR THE JOB

Make sentences with the phrases in the table.

Example: *A screwdriver is for tightening screws.*

A screwdriver	is for	cutting	bolts
A multimeter		making	components
A spanner		gripping	electrical circuits
A soldering iron		measuring	holes
A pair of wire strippers		joining	metal
A power drill		stripping	nuts
A vernier calliper		testing	objects
A pair of pliers		tightening	screws
A hacksaw		turning	wires
An Allen key		loosening	

11 ORDERING TOOLS

5

Ben Clark needs a new vernier calliper. He phones TechTools to order one.

a Look at the sentences below. Can you put them in the correct order to make a dialogue? (Tip: start with 13.)

b Listen to the CD to check your answers.

1	Ben	Can you invoice my company?
2	Ben	Do you have any other vernier callipers?
3	Ben	Hello, I'd like to place an order.
4	Ben	How much is that?
5	Ben	I'd like to order a 300 mm vernier calliper, please.
6	Ben	I'll have one of those, please.
7	Ben	That's OK. Cash on delivery, then, please.
8	TechTools	How would you like to pay?
9	TechTools	I'm afraid we don't have any 300 mm vernier callipers in stock.
10	TechTools	I'm sorry. We only offer credit card payments or cash on delivery to new customers.
11	TechTools	It's forty-eight euros ninety-five.
12	TechTools	Of course. What would you like, sir?
13	TechTools	TechTools, Lee speaking. How can I help you?
14	TechTools	Yes, we have a 200 mm vernier calliper.

12 PAIRWORK: ORDERING TOOLS

Work with a partner.

Student A: Turn to file 1 on page 126. **Student B:** Turn to file 10 on page 129.

GRAMMAR

The simple past

1 I **specialized** in panel assembly.
2 Last week I **did** a job in a block of flats.
3 Last week I **worked** nights.
4 What time **did** you **start** work?
5 I **didn't get** home until 10 a.m.

- Das *Simple past* benutzt man für abgeschlossene Handlungen in der Vergangenheit. Signalwörter: *Yesterday, last week, two days ago, When...?*
- Es wird mit der 2. Form des Verbs gebildet.
- Viele Verben haben eine unregelmäßige 2. Form. Sehen Sie sich in Zweifelsfällen die Liste auf Seite 189 an. (2)
- Fragen werden mit *did/didn't* + Grundform des Verbs gebildet. (4)
- Verneinungen werden mit *didn't* + Grundform des Verbs gebildet. (5)

13 PRACTICE

Electrician Ben Clark is talking about his training. Complete the text with the correct (simple past) forms.

I ... (leave)[1] school at 16. I ... (not know)[2] what to do, but my teacher ... (say)[3] that electrical engineering ... (be)[4] a good career. I ... (start)[5] an apprenticeship with a local electrical firm. During my apprenticeship I ... (learn)[6] practical skills on the job and also ... (go)[7] to college on day release. After three years I ... (pass)[8] my NVQ Level 3 in Electrical Panel Building and ... (become)[9] a qualified electrician. I ... (not want)[10] to start work straight away, so I ... (spend)[11] a year travelling around the world, then I ... (get)[12] the job that I do now.

14 REPAIRING A DEVICE

6

Waitz Rollenlager GmbH have purchased a grinding machine from Millers plc, but the machine has developed a fault. Millers send out a technician, Mike Oldman, to fix the fault.

Read or listen to the dialogue on the next page, then answer the questions below.

1 What is the problem?
2 How does Mike find the fault?
3 What action does he take?

MIKE Hi, I'm Mike Oldman, from Millers in the UK …
JAN Jan Kerschner. Pleased to meet you.
MIKE You sent me the repair message that the grinding machine for your roller bearing production has stopped working.
JAN Yes, that's right. The machine's motor went off and I couldn't restart it.
MIKE Well, let's have a look at the unit … Ah, it looks as if the contacts of the contactor are burnt. The contactor is designed to break high current load and serves as a kind of security gate for the machine. If it's faulty, the whole machine doesn't work. We've got to replace it.
JAN We've some spare contactors in stock. I'll get one.

Jan brings the new contactor and Mike installs it.

MIKE OK, let's see what happens now. Can you start the machine, please?

Jan presses the start button, but nothing happens.

JAN It still doesn't work.
MIKE So there must be another problem in the system. I'll check the inverter which controls the rotating speed of the motor. If the inverter doesn't work, the motor won't run.

Mike tests the power output from the inverter.

MIKE Right, I'm not getting a reading on the multimeter, so it looks like we've got a broken inverter. I'll have to install a new one.

He installs the new inverter, then Jan runs a function test.

MIKE The motor's working again. But it isn't running smoothly …
JAN It's getting hot, too.
MIKE (*Touches the motor*) You're right. We'll have to check the motor … Well, the multimeter shows a low ohm reading for the armature winding resistance of the motor. That's why the motor got overloaded, and this fault probably caused all the problems in the system.
JAN Can you repair the motor?
MIKE I don't think it's worth repairing, we'll have to replace it. Sorry!

15 WRITING A REPORT

Schreiben Sie einen Schadensbericht über die Reparatur auf Deutsch.

16 IT'S YOUR TURN

Work with a partner:

Student A: Give an example of a repair you have done in your job. Describe the problem and explain how you fixed it.
Student B: Ask questions about the repair. Make notes in English.

17 DIALOGUE

Work with your partner. Use your notes from exercise 16 and the zigzag diagram to make a dialogue. The expressions in the Language box may help you.

Electrician	Machine operator
Begrüßen Sie den/die Mitarbeiter/in und stellen Sie sich vor.	
	Stellen Sie sich ebenfalls vor.
Bitten Sie ihn/sie darum, das Problem zu beschreiben.	
	Beschreiben Sie das Problem.
Fragen Sie, wann das Problem aufgetreten ist.	
	Antworten Sie.
Erklären Sie, dass Sie die Maschine jetzt testen werden.	
	Sie sind damit einverstanden.
Sie haben die Ursache des Problems gefunden. Erklären Sie kurz die Ursache.	
	Fragen Sie, ob die Maschine sich reparieren lässt.
Antworten Sie darauf.	

LANGUAGE

Fixing a fault

Asking for information
What seems to be the problem?
When did the problem first occur?

Explaining what you are going to do
Let's have a look at the …
I'm going to take a look at the …
I'll have to check the …

Explaining what the problem is
It looks like the … is broken/faulty.
You need a new …
I'll have to install a new …
We'll have to replace the unit.

ELECTRICAL APPRENTICESHIPS IN BRITAIN

>lsc
Leading learning and skills

Young people in Britain who want to become an electrician can take part in the national apprenticeship scheme. An apprenticeship combines paid work and study and is usually open to people between the ages of 16 and 24.

Young people who are looking for an apprenticeship can call the national apprenticeship helpline or their local Connexions office: both of these services will help them to find a suitable employer. Alternatively, they can get in touch directly with a local employer.

Electrical apprenticeships are available in power generation, electrical distribution and supply and electrical and electronic servicing. They normally take about four years to complete.

Apprentices are given time off work to go to a local college. The exact balance of work and study depends on the employer. The study time may be one or more days each week (day release) or several weeks at a time (block release).

A basic wage or training allowance is guaranteed during the apprenticeship, and apprentices receive at least 20 days holiday per year.

Participants who finish their apprenticeship don't just get one qualification: they get a package of qualifications:

- an NVQ (National Vocational Qualification)
- a BTEC National Diploma or other technical certificate
- a Key Skills qualification which shows that they have important work skills like teamwork, problem solving, communication and using new technologies

Just like in Germany, doing an apprenticeship is hard work and a big commitment, but the rewards are good: there is a lot of demand for qualified electricians in Britain.

WORKING WITH THE TEXT

1 **Inform your classmates how to become an electrician in Britain by giving an outline of the text in German.**

2 **Compare the British apprenticeship scheme to the system in your country.**

UNIT 3
House wiring, 1

1 WARM UP

Work with a partner. Match the words in the vocabulary box with the numbers 1–12 in the diagram.

circuit breakers ■ conduit ■ consumer unit
earth terminal ■ electricity meter ■ flex
junction box ■ light switch ■ plug
RCD (residual current device)
service entrance ■ socket

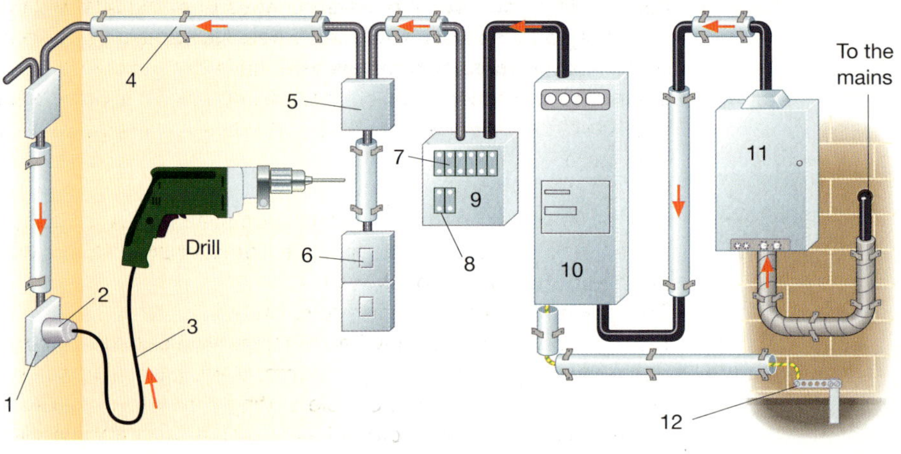

2 MEETING A CUSTOMER

7

Trevor and Penny Bell are English website designers. They have just bought an old house near Essen and they want some more sockets for their computers and other equipment. They contact Elektro Decker, an electrical contractor in Essen. Frank Böhm from Elektro Decker goes out to meet the Bells and discuss what they need.

Read or listen to the dialogue, then complete the exercise on the next page.

Frank rings the doorbell. Trevor opens the door.

TREVOR Hello?
FRANK Guten Tag. Mein Name ist Frank Böhm von der Firma Elektro Decker. Sind Sie Mr Bell?
TREVOR I'm sorry, I don't speak much German …
FRANK That's OK, sir. My name's Frank Böhm from Elektro Decker. Are you Mr Bell?
TREVOR Yes, I am. Pleased to meet you.
FRANK Pleased to meet you. (*They shake hands.*) You phoned us yesterday to ask about some changes to the wiring.
TREVOR Yes, that's right. Come in, please.

They go into the living room.

TREVOR This is my wife, Penny.
FRANK Pleased to meet you, Mrs Bell.
PENNY Pleased to meet you.
TREVOR We've just bought this house, so we want some changes, including a lot more sockets. In most rooms there's only a light switch and one socket.
PENNY And the kitchen's a disaster!
FRANK How do you mean?
PENNY We want a modern cooker and a microwave but there's nowhere to attach a cooker and not enough sockets for the kitchen appliances.
TREVOR That's because they had a gas cooker here and we want an electric one. We also want internet and satellite TV.
FRANK I see. As this is an old house let me just unscrew this light switch to see what sort of wiring you have and if you have channels or conduits in the walls … (*He unscrews a light switch.*) OK. You've got conduits, which is good, but I can see that you have the old-style wires from the 50s and 60s. In those days they only had black and red wires. Nowadays we have black, blue and green-yellow wires. This means we'll have to put new wiring in all the rooms where you want extra sockets because we aren't allowed to use the old black and red wiring any more. But the worst thing for you is the missing earth wire.
PENNY What do you mean?
FRANK Your house wiring doesn't have a protective earth conductor, Mrs Bell, so if there's a fault in a kitchen appliance, for example, the current won't travel to earth – it will travel through you!
PENNY That's awful! Did they just forget to put it in?
FRANK No, they didn't forget. They had different standards in those days. Nowadays we earth everything – baths, showers, water pipes and so on. There have been quite a few cases of people getting a shock off a water pipe in an old house because the pipe didn't have a protective earth conductor.
TREVOR I see what you mean.
FRANK So, to make your house really safe, I suggest we rewire it completely. We'll also need to put in a new circuit if you want an electric cooker. As I said, if we rewire the whole house, it'll be much safer for you. Now let's take a look at the consumer unit. Is it in the cellar?
TREVOR The consumer unit? Ah, yes that's what they call the fuse box nowadays. Yes, it's in the cellar. Come this way, please.

Are the following statements are true, false or not mentioned in the text? Correct the false statements.

1 Penny phoned Elektro Decker.
2 The kitchen is the biggest problem.
3 Penny and Trevor want a modern cooker.
4 Trevor shows Frank the wiring.
5 In the 60s they had wires of three colours.
6 Today we use more than two wires and more than two colours.
7 One of the people who lived in the house before Penny and Trevor got a shock off a water pipe.
8 Penny and Trevor only need a new consumer unit.

3 WORKING WITH WORDS

Read the dialogue again. Find the English translations for the following German words.

1 Elektroinstallationen, Verkabelung
2 Steckdose
3 herausschrauben
4 Leerrohr, Elektroinstallationsrohr
5 Schutzleiter
6 Strom
7 Stromkreis
8 Sicherungskasten

INFORMATION

Cables und *flexes*

Im Englischen bedeutet *cable* „fest installierte/s Kabel/Leitung", z.B. auf oder unter Putz. Flex (auch *cord* oder *lead* genannt) bedeutet „Netzkabel". Flex ist die Abkürzung von *flexible* = biegsam, elastisch.

4 CABLE OR FLEX?

Which of the following pictures show a flex and which show a cable?

1
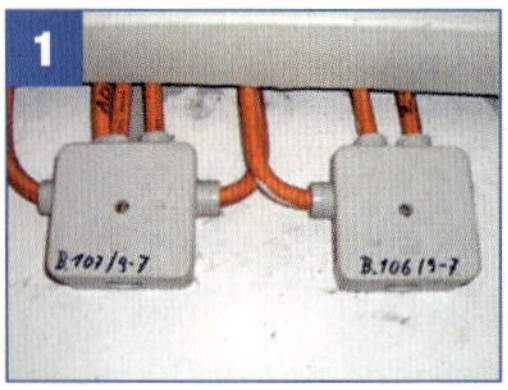

2

3

4

5

6
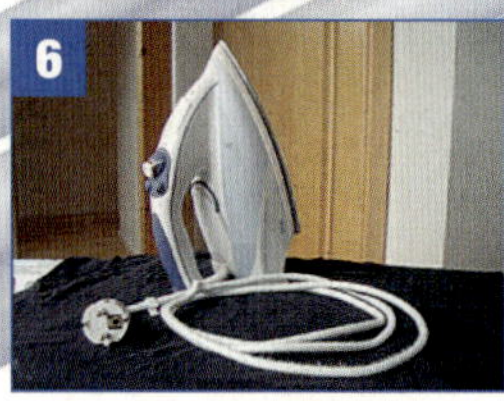

7
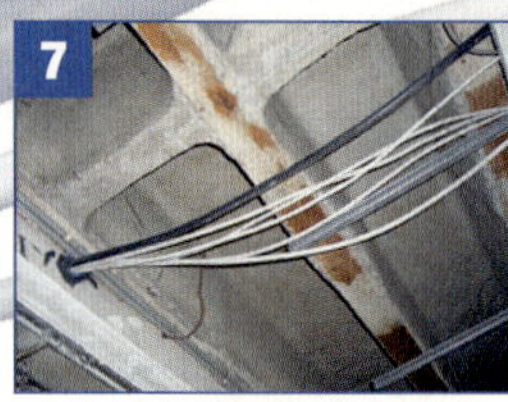

8

5 HOUSE WIRING PAST AND PRESENT

Penny and Trevor are talking about the electrician's visit. Complete the dialogue with the simple present or simple past forms of the verbs in brackets.

PENNY What … (the electrician / say)[1] about black and red wires, Trevor?

TREVOR He … (say)[2] that in the 60s old houses … (have)[3] black and red wires.

PENNY That's right. Nowadays houses … (have)[4] three-colour wiring: black, blue and green-and-yellow.

TREVOR Years ago they … (not have)[5] the third conductor. What … (the electrician / call)[6] it?

PENNY I … (remember)[7]. He … (call)[8] it the 'protective earth conductor'. It … (be)[9] green-and-yellow and it … (earth)[10] baths, showers and water pipes.

6 DESCRIBING A MODERN ELECTRICAL FITTING

Jeff Taylor, a young trainee electrician from England, is helping Frank Böhm to install some two-way light switches.

Look at the diagram, then put Frank's instructions in the right order. The first instruction is 'b'.

a Connect the conductors to the terminals of the light switch.

b Switch off the circuit breaker for this circuit, lock it and tag it.

c Cut away about 16 mm of insulation at the end of each of the three conductors.

d Cut the conductors to the right length.

e Push the switch into the mounting box in the recess in the wall and tighten the two big screws.

f Put on the plastic cover.

g Run the new cable through the conduit.

h Select the three conductors you will need for the light switch.

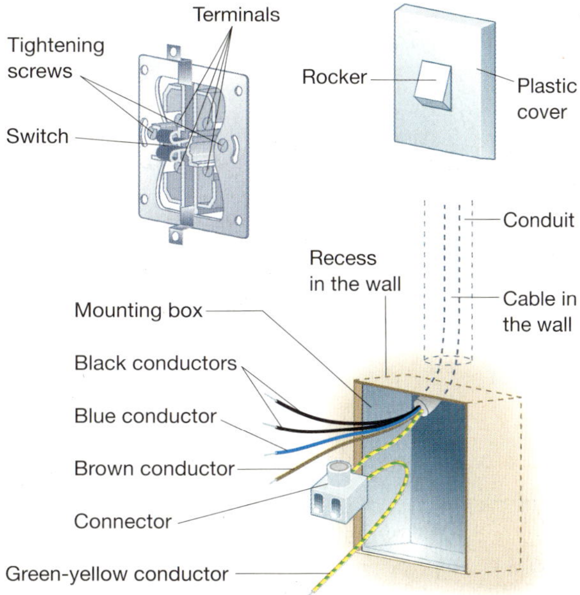

7 NAMING COMPONENTS

With the help of the diagram of the two-way switch, name the parts of the socket.

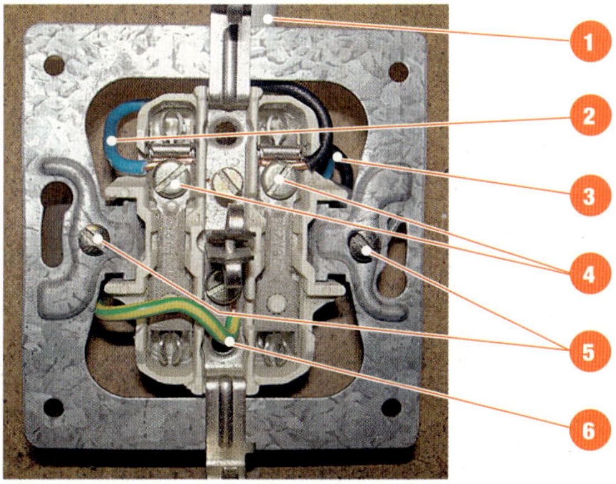

8 WIRING INFORMATION

Complete the information about house wiring with the words in the box.

circuit breaker ■ circuits ■ conduits
current ■ fuse ■ protective earth
rewire ■ socket

ELECTRICITY IN THE HOME

The electricity to power a machine normally comes out of a ...[1] in the wall. The cables in the wall often run through pipes known as ...[2] and have three conductors. The green and yellow conductor is called the ...[3] conductor. If there is an earth fault, the ...[4] 'escapes' along this conductor and not, for example, through a person's body, so the person is safe. Sometimes it's best to ...[5] an old house, if it only has black and red wiring. In a modern house there are separate ...[6] for each floor and for the bathroom and other equipment, too. If too much electricity passes through a wire, a ...[7] or a ...[8] in the consumer unit switches off the supply.

9 NEW WIRING FOR THE KITCHEN

8

Frank Böhm comes to talk to Penny about the wiring for the kitchen.

Listen to the dialogue, then choose the correct statements (a or b).

1 a The people in the house before Penny and Trevor had an old electric cooker.
  b The people in the house before Penny and Trevor had a gas cooker.
2 a Penny and Trevor want their new cooker near the window.
  b Penny and Trevor want their new cooker near the door.
3 a The fridge-freezer combination will be on the right.
  b The fridge-freezer combination will be on the left.
4 a There will be a double socket in the kitchen.
  b There won't be a double socket in the kitchen.

10 SPOT THE MISTAKES

8

Frank's plan for the work has three mistakes. Listen again and find them on the plan. (Refer to the list of symbols opposite as necessary.)

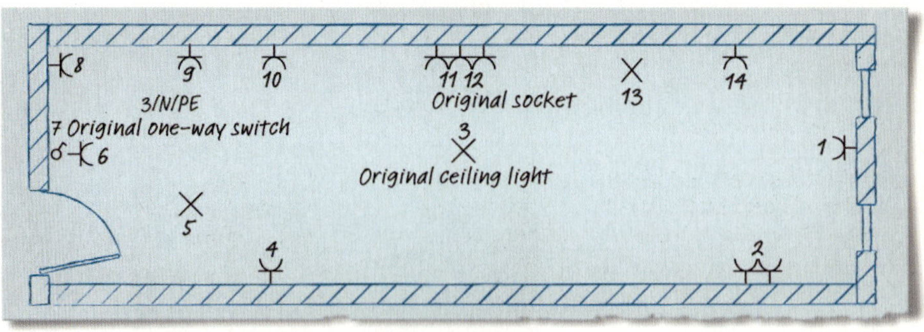

11 NEW WIRING FOR THE MAIN BEDROOM

Work with a partner to plan the new wiring for the main bedroom.

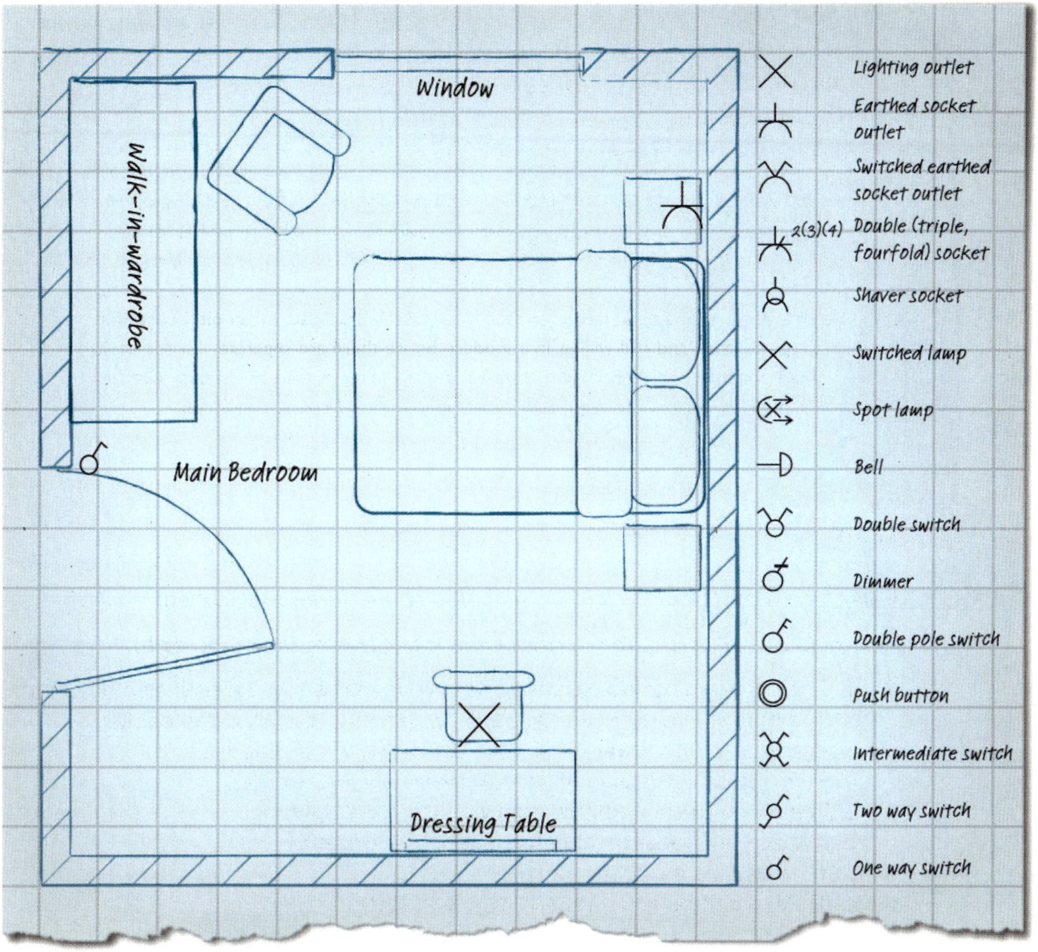

1 Look at the diagram and decide where to put new sockets, lighting outlets and other fittings.
2 Copy the plan and mark the new fittings on it, using the list of symbols.
3 Present your results to the class and explain your decisions. Use the Language box to help you with your presentation.

LANGUAGE

Describing where to install a fitting

We'll install a/an … for the …

to the left/right of the (armchair)	in the corner	by the (door)
on the left-hand / right-hand wall	over the (bed)	near the (dressing table)
in the middle	under the (window)	next to the (walk-in wardrobe)

CULTURE

Wie man Kunden anredet

Englischsprachige Kundinnen und Kunden erwarten, dass man sich mit beiden Namen (Vor- und Nachnamen) oder nur mit Vornamen vorstellt. Danach sagt der Kunde / die Kundin oft *Pleased to meet you,* worauf man wieder *Pleased to meet you* antwortet.

Um einen Kunden / eine Kundin höflich anzusprechen, benutzt man *sir* bzw. *madam*, z.B. *Would you like the socket here, madam?*

Männer können auch mit *Mr* + Nachname angesprochen werden, z.B. *Mr Bell.*

Bei Frauen ist es nicht so einfach, weil man *Mrs* (für eine verheiratete Frau), *Miss* (für eine unverheiratete Frau) und auch *Ms* zur Auswahl hat. Letzteres ist eine neutrale Form der Anrede von Frauen, die meistens nur geschrieben wird.

Tipp: Benutzen Sie bei Frauen das Wort *madam*, es sei denn, Sie sind sicher, dass die Kundin verheiratet ist.

12 INTRODUCING YOURSELF

Work with a partner to role-play these three dialogues. Use the Language box below.

Dialogue 1

Your first customer is Mrs McTavish, an elderly Scottish lady. Introduce yourself and your colleague Marco Trojahn.

Electrician	Mrs McTavish
	Sie öffnen die Haustür und sagen *Yes?*
Sie stellen sich und Ihren Kollegen höflich vor. Sie sollen die Waschmaschine reparieren.	
	Sie antworten höflich und stellen sich vor.
Sie antworten höflich.	

LANGUAGE

Introducing yourself to a new customer

Informal: Hi! I'm Frank from Elektro Decker and this is Silke.
Neutral: Hello. My name's Frank Böhm from Elektro Decker and this is Silke Braun.
Polite: Good morning/afternoon/evening, sir/madam. My name's Frank Böhm from Elektro Decker and this is my colleague Silke Braun.

Only use the informal version if you're sure that it's appropriate, e.g. with young people / students.

Dialogue 2

Your next customer is a young English student.

Electrician	Steve
	Sie öffnen die Haustür und sagen *Hi!*
Sie stellen sich und Ihren Kollegen umgangssprachlich vor.	
	Sie sagen nochmals *Hi!* und nennen Ihren Vornamen.
Sie erklären, dass Sie die Sicherungsautomaten überprüfen sollen. Sie bitten Steve, Ihnen den Sicherungskasten zu zeigen.	
	Sie antworten höflich und sagen, dass er im Keller ist. Sie bitten die Besucher, Ihnen zu folgen.

Dialogue 3

Your final customer this morning is an American businesswoman.

Electrician	Fiona Johnson
	Sie öffnen die Haustür und sagen *Hello?*
Sie stellen sich und Ihren Kollegen vor. Sie sollen den neuen Elektroherd anschließen.	
	Sie antworten freundlich und stellen sich vor. Sie geben den Besuchern die Hand und benutzen die entsprechende Begrüßungsformel.
Sie antworten entsprechend.	
	Sie bitten die Besucher, Ihnen in die Küche zu folgen.

LEARNING

Wie man Sprache für den Berufsalltag lernt

Neue Ausdrücke und Formulierungen lassen sich auf verschiedene Arten lernen, etwa indem man die einzelnen Wörter „paukt" oder sich die entsprechende Grammatik aneignet. Manchmal hilft es jedoch, nicht auf einzelne Vokabeln und grammatische Regeln zu achten, sondern sich gebräuchliche Wendungen als ganze einzuprägen. Denken Sie an Reisesprachführer, in denen Touristen praktische Wendungen an die Hand gegeben werden, mit denen sie sich im Urlaub „durchschlagen" können.

Denken Sie an diese Methode, wenn Sie eine Fremdsprache für Ihren Beruf lernen. Dort sind feste Wendungen, die man immer und immer wieder braucht, besonders wichtig. Der Erfolg stellt sich schneller ein, wenn man eine Frage wie *What seems to be the problem, sir?* nicht aus einzelnen Vokabeln und der grammatischen Frageform zusammenbaut, sondern sie sich direkt vollständig einprägt.

Schreiben Sie sich immer nützliche Wendungen auf, wenn Sie ihnen begegnen, und notieren Sie dabei, um welche Situation es jeweils geht.

GRAMMAR

If-sentences type 1

1 If there**'s** a fault in a kitchen appliance, the current **won't travel** to earth – it **will travel** through you!
2 If we **rewire** the whole house, it**'ll be** much safer for you.
3 We**'ll need** to put in a new circuit if you **want** an electric cooker.

- Ein *if*-Satz des Typs 1 drückt eine Bedingung aus, die der Sprecher für durchaus möglich oder wahrscheinlich hält. Der damit verbundene Hauptsatz drückt eine Vorhersage aus, die je nach Situation als Warnung, Versprechen o.ä zu verstehen ist.
- Der *if*-Nebensatz steht im *Simple present*, im Hauptsatz steht das *will-future*.
- Steht der *if*-Nebensatz an erster Stelle, wird er vom Hauptsatz durch ein Komma getrennt. (1, 2)

NB „If plus will makes me ill!"
If we ~~will~~ rewire the whole house …

13 DANGEROUS SITUATIONS

Complete the warnings about what will happen by using the correct forms of the verbs in brackets.

1 If the workman (touch) the overhead power cable with the ladder, he (get) a shock.
2 He (kill) himself if he (not be) careful.

3 The young man (have) a nasty accident if he (not tie) his hair back.
4 If he (drill) through the cable, the circuit breaker (trip).

5 The circuit breaker in the cellar (not trip) fast enough to save the man if the radio (fall) into the water.
6 If the man (not take) his hand out of the water, he (electrocute) himself.

7 If Billy (turn) on the light, he (give) his father a dangerous electric shock.
8 The microwave (fall) onto his head if Billy (pull) the flex.

EXTRA MATERIAL

FROM THE POWER STATION TO THE CONSUMER

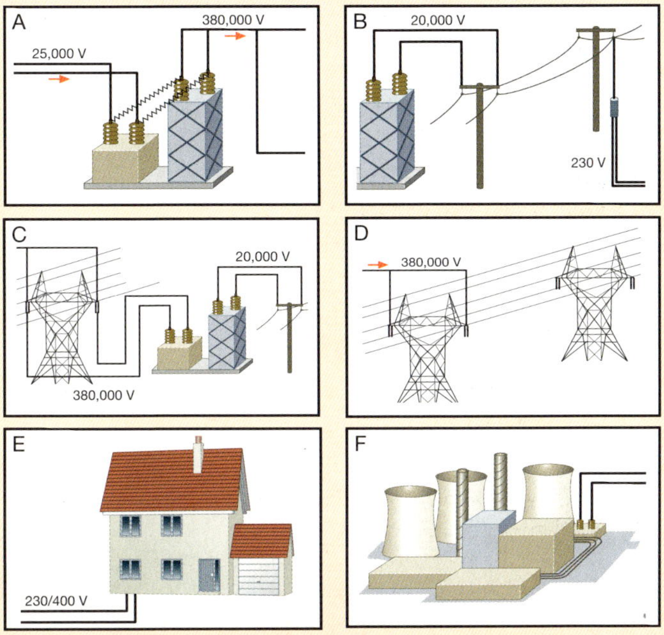

- Finally, it is taken to the consumer, often via underground cables.
- It is then transported overland at 380,000 V along power cables held up by electricity pylons.
- Next it is stepped down again to 230/400 V.
- Next it is fed into the national grid and the voltage is stepped up at a substation by a transformer.
- First electricity is generated at a power station.
- Then the voltage is stepped down to 20,000 V.

1 EXPLAINING ELECTRICITY SUPPLY

Write the bullet points in the correct order to explain how electricity gets from the power station to the consumer. Then find the picture which goes with each point.

Example: *First electricity is generated at a power station (F).*

2 GOING INTO DETAIL

Explain why the voltage is stepped up and down.

UNIT 4
House wiring, 2

1 WARM UP

a Which of the following signs …

- are the strongest warnings?
- are strong warnings about permanent danger?
- are general rules and regulations?

b Explain in your own words what each sign means.

c Give an example of where each sign is used.

d When did you last see one?

2 REWIRING THE GARAGE

9

Trevor and Penny Bell's old house has now been rewired. Today they're in the living room talking to Frank Böhm from Elektro Decker about another project.

Read or listen to the dialogue, then answer the questions below.

1 What do Trevor and Penny want to do with the garage?
2 How has Trevor supplied power to the garage?
3 Why is this a problem?
4 Explain why Trevor and Penny need a water heater in the garage.
5 What kind of heater does Frank recommend, and why?

TREVOR Thanks for coming so quickly, Mr Böhm. We'd like to discuss our plans for the garage with you.

FRANK Yes, my boss tells me you want electricity in the garage so you can use it as an apartment.

PENNY Yes, that's right and we don't want to see all those ugly cables. Please come this way …

Outside the garage:

TREVOR The lights and sockets in the garage don't work, so I've got an extension from the house to the garage.

FRANK Well, I wouldn't do that if I were you, I'd take it down because there could be a nasty accident. You'll need an underground cable from the house to the garage. Anything else would be too dangerous.

PENNY I'd take his advice if I were you, Trevor. You could get an electric shock if you touched it in the rain!

TREVOR All right. I'll take it down in a minute – but let's go in the garage first and show Mr Böhm what we want.

Inside the garage:

TREVOR We want to put in a kitchenette and a shower.

FRANK Oh, I see. A shower. In that case you'll need a strong water heater.

PENNY Will we need an immersion heater?

FRANK You mean one with a hot water tank?

PENNY Yes, that's right.

FRANK No, I wouldn't install an immersion heater if I were you. If you put a big tank on the wall, it would take up too much space in a small place like this. What you need is a continuous flow water heater.

TREVOR So that's a tankless heater, isn't it?

FRANK Exactly. They're efficient and compact and it won't be a problem to wire the garage for one. You'll need 400-volt three-phase current and a small consumer unit with circuit breakers and an RCD to be on the safe side.

PENNY OK. I've got a sketch here of where we want the light switches, sockets, shower and cooker.

FRANK That's very useful. I'll use it to see what materials we'll need and give you an estimate.

PENNY Good!

CULTURE

An Englishman's home is his castle

Briten finden es seltsam, dass so viele Deutsche mieten, anstatt ein eigenes Haus oder eine eigene Wohnung zu kaufen. Ca. 70 Prozent der Bevölkerung in Großbritannien wohnt in einem Eigenheim (in Deutschland sind es 40 Prozent). Hat man das erste Eigenheim gerade mittels einer Hypothek (*mortgage*) erworben, plant man bereits den nächsten Umzug, um ein teureres, größeres Haus zu kaufen. Das nennen die Immobilienmakler *climbing the property ladder*. Dies hat auch zur Folge, dass viele Briten begeisterte Heimwerker sind. Dieses Hobby heißt auf Englisch *DIY* (*Do it yourself*). Unter den *DIYers* gibt es natürlich auch welche, die sich bauliche Änderungen und auch elektrische Installationen zutrauen, für die ihre Fachkenntnisse gar nicht ausreichen …

3 PLANNING THE JOB

Work with a partner.

a **Make a list in German of the materials and equipment Frank will need.**

Example: *Durchlauferhitzer*

b **Trevor often works outside doing repairs to the house, working in the garden, cutting the hedge and mowing the lawn. What electrical fixture would you recommend for his power tools?**

GRAMMAR

If-sentences type 2

1 I **wouldn't do** that if I **were** you.
2 If you **put** a big tank on the wall, it **would take up** too much space.
3 If you **had** more money, **would** you **buy** a bigger house?

- Durch die Wahl eines *if*-Satzes des Typs 2 gibt der Sprecher seinem Gesprächspartner einen Ratschlag oder teilt ihm mit, dass er die Erfüllung der Bedingung für unmöglich, unwahrscheinlich oder nicht wünschenswert hält.
- Der *if*-Nebensatz steht im *Simple past*, im Hauptsatz wird *would* + Infinitiv verwendet.
- Steht der *if*-Nebensatz an erster Stelle, wird er durch ein Komma vom Hauptsatz abgetrennt. (2, 3)

NB „If plus would is no good!"
~~If I would~~ …

4 PRACTICE

Frank Böhm is talking to Jeff Bridges about the wiring system in the converted garage.

Choose the correct verb forms to complete the dialogue.

JEFF If Trevor and Penny … (want)[1] to put in a 230 V undersink water heater instead of the tankless one, … (it/be)[2] a problem?

FRANK No, but if we … (have to)[3] replace a 230 V water heater with a 400 V model, we … (have to)[4] change the wiring because of the three-phase current.

JEFF Oh, I see. If I … (have)[5] a garage like this I … (not convert)[6] it into an apartment, I … (use)[7] it as a radio shack.

FRANK That's because you haven't got a car. If you … (have)[8] a car you … (park)[9] it in your garage!

5 WHAT WOULD HAPPEN IF ...?

Describe the following 'chain reaction' using if-sentences.

Example: *If Frank used the wrong cable, the insulation would melt.*
If the insulation melted, ...

Frank / use / the wrong cable → the insulation / melt → it / trip / the circuit breaker → the circuit breaker / switch off / the circuit → the lights and sockets / not work → Penny and Trevor / be / annoyed → they / call / the company → Frank / get / into trouble

6 IT'S YOUR TURN

Work with a partner. Invent two more 'chain reactions'.

Example: *If I didn't install an RCD in a bathroom circuit, ...*

7 READING A TECHNICAL TEXT

Find the paragraphs in the text which deal with the following subjects.

1 Conductors 2 Dangers 3 Safety devices

ELECTRICAL HANDBOOK

MODERN SAFETY STANDARDS

1 In an electrical circuit the current from the power station flows along the **hot wire** (phase – brown conductor) through the **load** (e.g. a light) and then along the **neutral wire** (blue conductor) back to the power source. The load in the circuit acts as **resistance** and stops an uncontrolled amount of current (a surge) from flowing along the hot wire directly back to the source.

2 If there is a fault in the circuit, e.g. if the brown and blue conductors touch because the insulation has melted (short-circuit), a massive amount of current suddenly flows through the circuit because there is almost no resistance in the circuit. This is a very dangerous situation because the extra energy takes the form of heat and can cause a fire.

3 For this reason a modern house wiring system includes safety devices to switch off the electricity supply if there is a surge in current.

4 There are basically two types of safety device in a modern home – **circuit breakers** to protect the circuit and **residual current devices** (RCDs) to protect people. A circuit breaker reacts if too much current flows through a circuit, whereas an RCD reacts if there is current leakage, e.g. in the case of a ground fault.

Figure 1: *circuit breaker*

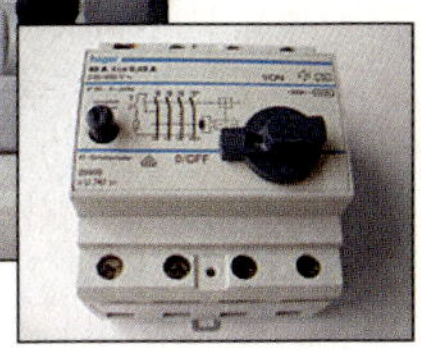
Figure 2: *RCD*

In a modern house or flat there will normally 5
be 10–20 circuit breakers and one or two RCDs, e.g. in the electrical circuit of a bathroom. Whereas a 16 amp circuit breaker will trip within 0.4 seconds if the current rises to five times the nominal value (e.g. 5 × 16 A = 80 A), an RCD will trip within 0.2 seconds if there is a fault current of only 30 mA.

8 UNDERSTANDING THE TEXT

Read the text again, then answer the questions.

1 What would happen if there were a short-circuit in an electrical circuit without a safety device?
2 Which device in the circuit would react first to a fault current of 40 mA – a circuit breaker or an RCD?
3 Bob's bathroom circuit has a circuit breaker but it doesn't have an RCD. If Bob got an electric shock while he was in the shower, what would happen? What wouldn't happen?

9 WORKING WITH WORDS

Complete the sentences with words from the text in the correct form.

1 Electricity moves like the water in a river. It … .
2 A material which lets electricity through easily is a good … .
3 A … in the circuit reduces the current.
4 A sudden increase in electricity is a … .
5 The … around the wires in a cable is normally a form of plastic.
6 When a solid turns into a liquid it … .
7 RCDs and circuit breakers are … .
8 RCDs … people and equipment from harm.

LEARNING

Wie man Vokabeln lernt

Zum Vokabellernen ist es ganz wichtig, ein Vokabelheft zu führen. Dort sollten Sie alle neuen Wörter mit deutscher Übersetzung eintragen.
Oft ist es hilfreich, sich statt einzelner Wörter gleich feste Ausdrücke und Wendungen einzuprägen. So wissen Sie gleich, wie das Wort im Satzzusammenhang verwendet wird:

trip – to trip a circuit breaker *einen Stromkreisunterbrecher auslösen*

10 GETTING YOUR FACTS RIGHT

Read the text on the next page, then complete the statements.

1 If five times the nominal current flows through a circuit …
2 The two different types of metal in a bimetallic strip …
3 The bimetallic strip breaks the circuit because ...
4 A solenoid is ...
5 A solenoid reacts more quickly ...
6 To complete the circuit again you ...

ELECTRICAL HANDBOOK

CIRCUIT BREAKERS

A circuit breaker breaks the circuit (trips) if the current flowing through it rises to an unacceptable level. This is usually between three and five times the current the circuit breaker is designed to take (nominal current).

Modern circuit breakers have a solenoid, a bimetallic strip or both.

Thermal-magnetic circuit breaker
In a thermal circuit breaker a bimetallic strip is made of two different types of metal which expand at different rates when heated. At room temperature the two strips of metal are both straight. If, however, a stronger current flows through them, heat is produced and the two metal strips start to expand. As one of them expands more than the other, it trips the electrical contact and breaks the circuit.

In the circuit breaker shown in Figure 3 there is a bimetallic strip on the right (3) and a solenoid in the centre (2). A solenoid is an electromagnetic switch. This combination has the advantage that a sudden large current surge will trip the solenoid, which reacts quickly (e.g. in 20 milliseconds), and a smaller long-term overload will trip the bimetallic strip, which reacts more slowly but to a lower current. The circuit can be completed again by pushing the switch (1) up by hand.

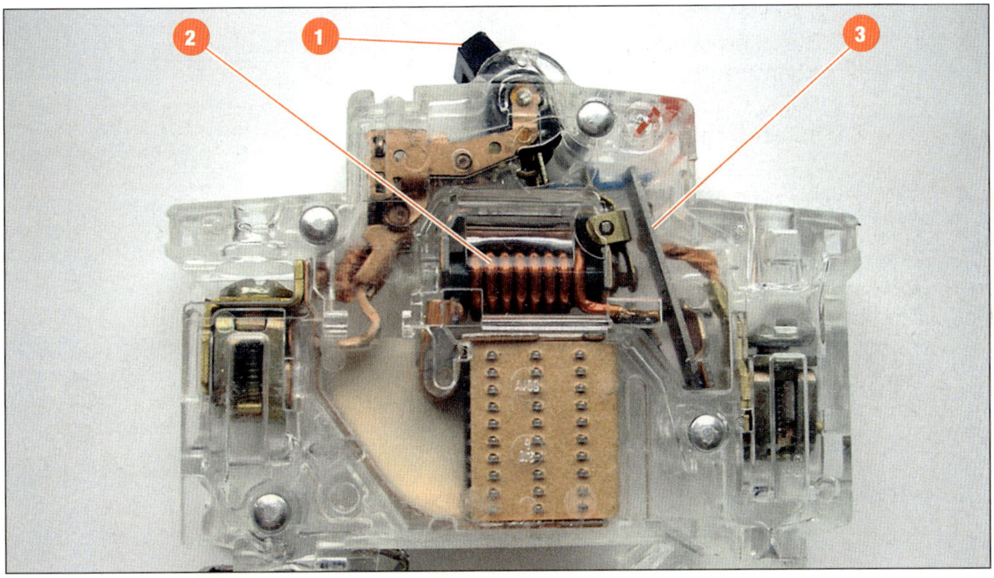

Figure 3: *Thermal-magnetic circuit breaker*

11 WORKING WITH WORDS

Match a–e with 1–5. The expressions are all in the text.

1 break	a the solenoid
2 produce	b the switch
3 bend	c the strip
4 trip	d heat
5 push up	e the circuit

12 MEDIATION

Summarize the text on page 41 in German for a colleague who does not speak English. Do not translate it word-for-word.

13 HOW AN RCD WORKS

Complete the text below with words from the diagram (figure 4).

ELECTRICAL HANDBOOK

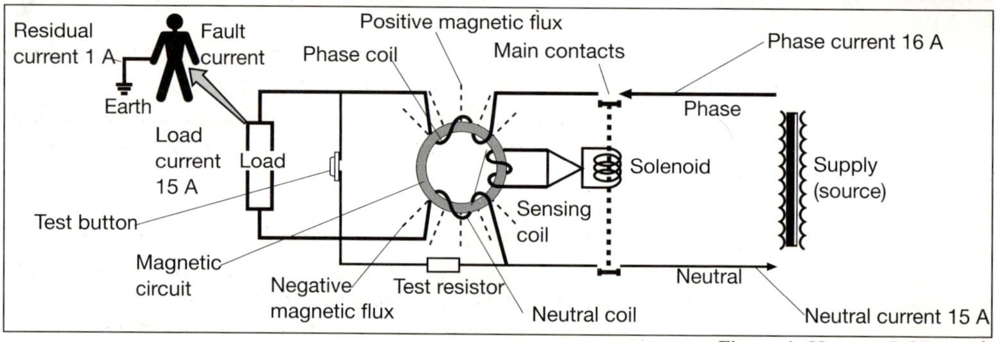

Figure 4: *How an RCD works*

RESIDUAL CURRENT DEVICE (RCD)

In figure 4 the …[1] current of 16 amps travels to the load in the left-hand part of the circuit and the …[2] current of 15 amps returns from it. As there is a …[3] current of 1 amp, the RCD switches off the circuit.

In a circuit without a fault, the current enters the phase …[4] on its way to the load and returns to the …[5] through the neutral coil. A magnetic …[6] is produced in both coils. However, as the coils are identical but wound in opposite directions, they cancel each other out because one is …[7] and the other …[8].

In the case of a fault, some current is lost to …[9] and the …[10] current returning to the supply is less than the …[11] current to the load. The …[12] flux in the neutral coil is therefore less than in the phase coil. This difference is picked up by a …[13] coil, which is connected to a …[14]. This disconnects the main …[15] and switches off the circuit.

14 TRUE OR FALSE?

Study the diagram and text and then say whether the following statements are true or false. Correct the false statements.

1. The residual current is the difference between the phase and neutral currents.
2. The phase coil and the neutral coil are wound in the same direction.
3. The current from the supply first produces a negative magnetic flux, then a positive one.
4. There is no physical difference between the phase coil and the neutral coil.
5. The test button is connected to the solenoid.
6. The sensing coil is wound around the neutral coil.
7. The sensing coil triggers the test button.

15 SAFETY AT WORK

When Jeff and Frank are in the workshop at Elektro Decker, Jeff sees a notice on the five safety rules. He asks Frank Böhm to help him understand them.

Complete the dialogue with German words from the notice.

> **5 Sicherheitsregeln**
>
> **Vor Beginn der Arbeiten**
> – freischalten
> – gegen Wiedereinschalten sichern
> – Spannungsfreiheit feststellen
> – erden und kurzschließen
> – benachbarte, unter Spannung stehende Teile abdecken oder abschranken

JEFF I'm not quite sure if I've understood this, Frank. Does *freischalten*[1] mean 'switch off'?

FRANK Yes, that's right, but it means more than that. It means you must disconnect an electrical circuit or a piece of equipment before you work on it.

JEFF Oh, I see. And what about …[2]? Has that got something to do with 'safety'?

FRANK Yes, but here it means 'protect' – protect the circuit against being turned on again.

JEFF OK, I get it. Now what does …[3] mean?

FRANK It means 'make certain'. We have to make certain the circuit isn't live – in German we say 'make certain it's *free of electricity*'. It's best to test it with a multimeter to make certain it's not connected to the mains.

JEFF I'm sorry to bother you but there's just one more thing. Could you explain …[4]?

FRANK Sure. It means 'earth and short-circuit'.

JEFF I won't even try the last one but I think …[5] has got something to do with 'neighbours', hasn't it?

FRANK No, you've got the wrong end of the stick there, Jeff. It means 'nearby', so the rule means 'Cover up or fence off nearby live components'.

JEFF OK, I've made a note of that.

16 CHECKING JEFF'S NOTES

Jeff made notes on the five safety rules. Check them and find the three mistakes he made.

1 *Switch off a piece of equipment before you work on it.*
2 *Tell everybody not to switch on the circuit again.*
3 *You can start work on a circuit after you have checked with a multimeter that it's not live.*
4 *Sometimes you need to earth and short-circuit an electrical circuit.*
5 *Make it impossible to touch live electrical components nearby if you can't switch them off.*

17 INSTALLATION INSTRUCTIONS

A German electrician is in Liverpool helping an English colleague to repair a shower. He/She wants to know what the installation instructions mean.

Work with a partner. Role-play the dialogue, using expressions from the Language box.

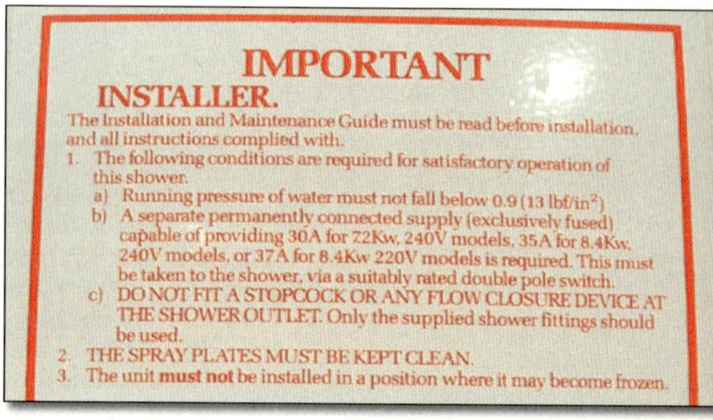

German electrician	English electrician
Sie möchten wissen, was *13 lbf/in*^{2} bedeutet.	
	Sie erklären, dass dies **pounds force per square inch** (Pfundkraft pro Quadratzoll) bedeutet. *lb* steht für *pounds* (Pfund) und *f* für *force* (Kraft). Der Wasserdruck ist hier gemeint.
Sie fragen, ob *exclusively fused* etwas mit einem Sicherungsautomaten zu tun hat.	
	Sie erklären, dass das Gerät einen eigenen Sicherungsautomaten haben muss.
Sie sind nicht sicher, ob Sie richtig verstanden haben. Bedeutet *permanently connected supply*, dass das Gerät direkt ans Stromnetz angeschlossen ist?	
	Sie bejahen. Man darf keine Steckdose benutzen.
Sie hoffen, dass Sie Ihrem Kollegen nicht auf die Nerven gehen und stellen eine letzte Frage. Wie hoch ist die Netzspannung in Liverpool?	
	Die Spannung war früher 240 V, ist aber jetzt 230 V.

LANGUAGE

Asking for information

What does … mean?
Does … mean … ?
Could you explain … ?
What about … ?

Has that got something to do with … ?
I'm not quite sure if I've understood this but …
I'm sorry to bother you but …
There's just one more thing …

HOW DANGEROUS IS AN ELECTRIC SHOCK?

Electricity can kill: most deaths result from a current entering the body and passing through the chest or the head. In all cases, the voltage (V) is less important than the strength of the current (A – amperage) and the duration (how long it flows through the body).

An electric shock of 1 mA isn't dangerous or painful to most people. You may feel a slight 'tingle', if anything. A 5 mA shock is unpleasant, but not usually dangerous.

With slightly higher currents, there is a danger that your muscles will 'freeze' and you won't be able to let go of the electricity source. This reaction greatly increases the duration of the shock and happens somewhere between 6 mA and 30 mA. Most adults can still let go at 15 mA AC (or 70 mA DC), and this value is known as the 'let go' or 'freezing' current. Higher currents than this are extremely painful and may cause death.

Currents of 1–4 amps disturb the rhythm of your heartbeat, causing ventricular fibrillation (uncoordinated heart function). Nerve damage is also likely. A shock of this strength may even stop your heart (cardiac arrest) and cause death.

A current of over 10 amps causes serious internal and external burns and will almost certainly cause death by cardiac arrest.

1 UNDERSTANDING THE TEXT

Read the safety information above, then copy and complete the table.

	Strength of AC current	Reactions
1	1 mA	*k*
2	5 mA	
3	6–30 mA	
4	50–150 mA	
5	1.0–4 A	
6	>10 A	

a cardiac arrest possible
b cardiac arrest likely
c death likely
d death possible
e extreme pain
f freezing of muscles
g mild pain
h strong pain
i nerve damage
j severe burns
k tingle
l ventricular fibrillation

2 FIRST AID

Explain what you should do if a colleague receives an electric shock.

Example: *1: Disconnect electricity source.*

UNIT 5
Installing a network

1 WARM UP

You are going to install a network for a small business. Note down questions which a customer might ask. Then answer the questions on your partner's list.

2 THE BENEFITS OF NETWORKING

Read the web page. Make notes on the advantages a network offers to a small business.

File Edit View Go Bookmarks Tools Help

http://www.smallbiznet.co.uk/index.html Go

What can a small business network do for me?

'My business doesn't need a network. Networks are for bigger companies. We're too small.'

Are you sure?

With stand-alone computers, work slows down when people need to share printers, or an internet connection, or information. Sure – you can buy more printers or modems, but that's expensive. You can email information to your colleagues, but that's an inefficient way of sharing large files. Backing up valuable data on stand-alone computers is time-consuming.

A small business network is the only realistic solution.

A network lets your people share printers and other peripherals – so you spend less money on hardware. It makes file sharing between colleagues easier. It allows one computer to backup the data of another – that makes your data more secure.

With a network, you can share one broadband internet connection among all your computers – that's more cost-effective.

A network makes your business more flexible, more efficient and more competitive.

Still think you don't need a network?

3 UNDERSTANDING THE TEXT

Complete these sentences about the text in your own words.

1 Some small businesses think they don't need a network because …
2 Just buying more peripherals isn't a good solution because …
3 If you have valuable data on stand-alone computers …
4 A network saves you money because …

4 TALKING TO A NEW CUSTOMER

10

NetzCom Berlin AG is a company specializing in computer networking. Robert King from King's English phones their hotline.

Read or listen to their dialogue. What is Robert King calling about? Give the main ideas in two or three sentences.

SCHELLER NetzCom, Peter Scheller. Was kann ich für Sie tun?
KING Oh, err … do you speak English? My German isn't so good …
SCHELLER Yes, I do. How can I help you?
KING I'm Robert King from King's English. I'd like to speak to someone in the … err … technical department, please.
SCHELLER Well, I'm in charge of networking projects …
KING Oh, good. I've got the right person, then … Well, we're a small private English school here in Berlin, and we're thinking of networking the computers in our office. Is that the sort of thing that you do?
SCHELLER Yes, of course.
KING Excellent. Well, as I say, we're thinking of networking our computers, but we aren't sure whether it's a good idea. I mean, we're only a small school, and we're worried about it being too expensive …
SCHELLER Well, usually a network saves our customers money in the long run, because they can share peripherals like printers and scanners and of course they can work much more efficiently.
KING Sounds good, but can you tell me how much it will cost?
SCHELLER Yes, I can, but I'll need more information about your requirements first, so that I can find the best networking solution for your situation.
KING Oh, yes, sure. What do you need to know?
SCHELLER Well, I need to know how many staff you have, what computers you have, whether you want us to network all of your computers, whether you want a wireless network and so on … Could you email me some details? A layout plan of your school would be useful as well.
KING OK, I'll do that this afternoon.
SCHELLER Great. When I've had a look at that, we can arrange a meeting. Now, let me give you my email address …

5 UNDERSTANDING THE TEXT

10

Read or listen to the conversation again. Decide whether the following statements are true, false or not in the text. Correct the false statements.

1 Robert King speaks a little bit of German.
2 Peter Scheller is in the sales department.
3 King's English wants to network the computers in its classrooms.
4 Eight teachers work in the school.
5 Robert is concerned about how much a network will cost.
6 The school wants a wireless network (WLAN).
7 Peter asks Robert to send him a layout plan of the school.
8 Peter would like to meet Robert to discuss the project.

6 WORKING WITH WORDS

Complete this advertisement for NetzCom with the words in the box.

arrange ■ efficiently ■ network
peripherals ■ requirements ■ share
solutions ■ staff ■ technical

NETZCOM

NETZCOM – networking …[1] for small businesses

At **NETZCOM**, we help you to …[2] your computers. That means that your …[3] will be able to …[4] printers and other …[5] and work more …[6]. Why not call our hotline to discuss your …[7]? Alternatively, one of our …[8] experts will …[9] a meeting with you.

7 DISCUSSION

Discuss in a small group:

You are going to install a network for a small business. What questions do you need to ask the customer at the start of the project? Make a list of the questions.

8 AN EMAIL FROM THE CUSTOMER

Sie sind Peter Schellers Assistent/in. Notieren Sie sich als Planungshilfe den Inhalt des Schreibens auf Deutsch.

Inbox - Outlook Express | School Network

!	0	⚑	From	Subject	Received

Hello Peter,
Here is some information about our school office to help you to plan the network. At the moment our school is fairly small. We have six teachers, a secretary and a receptionist. Our office equipment is basic: the secretary and the receptionist have a computer each, and the rest of us share the third computer in the office. The secretary's computer is the only one which has internet access.
There are two more computers in the two classrooms. Our students use these computers. They should be part of the network as well.
I also attach a sketch of the office which will show you where the staff work.
I hope this is enough information to go on. Please send me an email if you require more details.

Best wishes,
Robert

1 message(s), 0 unread | Working Online

9 WORKING WITH WORDS

Read the email again to find words or phrases with similar meanings. They are in the same order.

quite ■ simple ■ too ■ drawing ■ need

10 NETWORK TOPOLOGIES

King's English decide against installing a WLAN, but they want to know about all the other network options. Peter draws three different network topologies on the layout plan of the school.

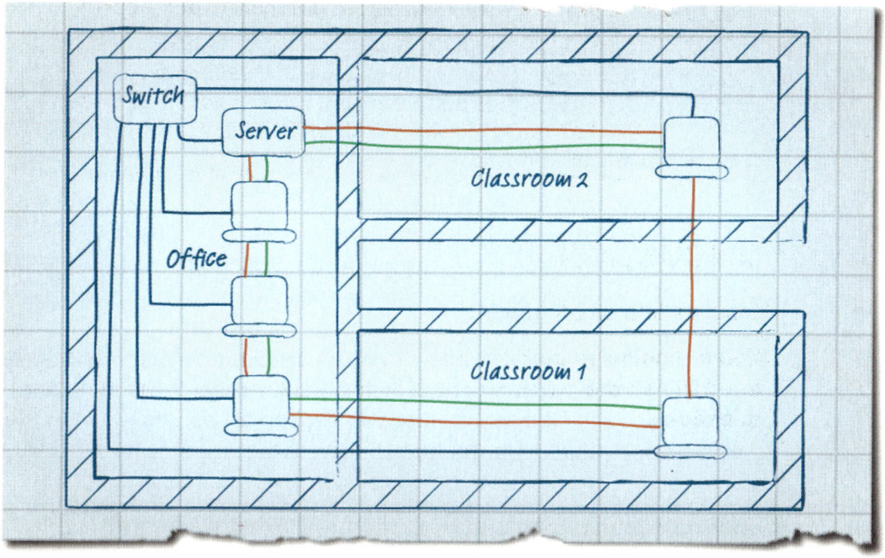

a Use the table to make sentences about the topologies. Then match them to the correct colours in the layout plan.

A	B	C
In a bus topology, In a ring topology, In a star topology,	the devices are connected	by separate cables to a central switch. to each another in a ring. to a main coaxial cable.

b Work with a partner. Decide which advantage(s) and disadvantage(s) (a–g) apply to each topology. Copy and complete the table.

Topology	Advantages	Disadvantages
Bus Ring Star	*e*	

a It can't support heavy traffic.
b It can support more traffic than the other networks.
c It's easy to add more workstations later.
d It's more expensive than the other networks.
e It's the simplest and cheapest network.
f If the switch is faulty, the system won't work.
g It's the easiest topology to manage.

c What other advantages and disadvantages can you think of?

11 ADVISING THE CUSTOMER

11

Peter Scheller discusses the configuration of the King's English LAN with Robert King.

Listen to their conversation. Does Peter Scheller mention the same advantages and disadvantages that you chose in exercise 10b?

12 WHAT DO THEY SAY?

11

Listen again. What do they say – a or b?

1 a Sorry, what's a 'LAN', exactly?
  b Sorry, I don't know what a 'LAN' is.
2 a Could you give me some more information about the networks, please?
  b Could you give me some information about the different possibilities, please?
3 a On the other hand, bus networks can't support heavy traffic.
  b The disadvantage of bus networks is that they can't support heavy traffic.
4 a I don't think that would be much good for us, then.
  b I don't think that would work for us, then.
5 a I don't think it's the best option for you.
  b I don't think it would be a good option for you.
6 a I suggest using a star topology.
  b I'd recommend a star topology.

LANGUAGE

Advising a customer

Making suggestions
I suggest …
I'd recommend …
What do you think of … ?

Accepting a suggestion
That sounds like a good idea.
That sounds sensible/fine.

Rejecting a suggestion
I don't think that would work.
I can see what you mean, but …

Asking for more information
Could you give me some information about … , please?
I'd like to know more about … , please.

Asking for clarification
Sorry, what's … , exactly?
I'm afraid I don't quite understand.

CULTURE

Höflichkeit

Englische Muttersprachler gebrauchen viele „überflüssige" Wörter, um einen höflichen und hilfreichen Eindruck beim Gesprächspartner zu machen. Schlichte Antworten können auf Englisch schroff und abweisend klingen.

Wenn Sie um etwas bitten, sollten Sie mit *Could you ...* anfangen und auf jeden Fall *please* hinzusetzen:
Could you call me back, please?

Vermeiden Sie es, eine Frage mit einem schlichten *Yes* oder *No* zu beantworten, und gebrauchen Sie bei passender Gelegenheit *thanks* bzw. *sorry*.
Did you get my email? – Yes, I did, thanks. *Can we meet this afternoon? – No, I'm afraid we can't. Sorry.*

Sagen Sie niemals *I want*, sondern *I would like*:
I'd like to speak to Peter Scheller, please.

13 BEING POLITE

Re-write the dialogue to make it more polite.

ALEX Budget Computers. What do you want?
TOM I want to speak to Lisa Smith.
ALEX You can't, because she's out this morning.
TOM Then write this down. I'm Tom Jackson. Tell Lisa to ring me. OK?
ALEX Yes.
TOM Bye.

14 ROLE PLAY: A NETWORKING PROJECT

Work with a partner.

Student A: Turn to file 2 on page 126.
Student B: Turn to file 11 on page 129.

When you have finished, swap roles.

GRAMMAR

Comparatives and superlatives of adjectives

1 It's the **simplest** and **cheapest** network, but not the **best**.
2 Networks are for **bigger** companies.
3 A network makes file sharing **easier**.
4 It's **more expensive than** the other networks.
5 You spend **less** money on hardware.
6 It isn't **as easy** to manage **as** the other networks.

- Alle einsilbigen Adjektive werden mit *-(e)r* und *-(e)st* gesteigert. (1, 2)
  Einzelkonsonanten nach einem kurzen Vokal werden dabei verdoppelt. (2)
- Zweisilbige Adjektive mit den Endungen *-y*, *-er*, und *-le* werden ebenfalls mit *-(e)r*, und *-(e)st* gesteigert. (1, 3)
  Dabei wird *-y* in *-i-* umgewandelt. (3)
- Alle anderen zwei- und mehrsilbigen Adjektive werden mit *more* und *most* gesteigert. (4)
- Unregelmäßig sind: *good–better–best; bad–worse–worst; little* (wenig) *–less–least.* (1, 5)
- Vergleiche bildet man mit *than* und *(not) as … as.* (4, 6)

15 PRACTICE

Complete these sentences from the website of a networking company. Use the comparative or superlative forms of the adjectives in brackets.

1 We will supply you with the … (up-to-date) information on networking.
2 Bus networks are the … (cheap) option.
3 Ring networks are … (expensive) than the other options.
4 Star topologies are the … (good) choice for networks at the moment.
5 They are the … (easy) networks to maintain.
6 Ethernet is the … (popular) network standard in the world.
7 Fibre optic cabling is … (advanced) than copper cabling.
8 A WLAN is the … (flexible) option but it has the … (big) security problems.

16 SETTING UP THE NETWORK

Peter and his assistant are discussing how to set up the new network.

Work with a partner:

a **Choose the six most important tasks from the list. Put them in chronological order.**

b **Now add any other tasks which you think are important.**

List of tasks to be carried out in the project
Write an invoice
Familiarize the customer with the system
Install and configure the operating system
Run the hardware and software test
Order the parts from the supplier
Do the network cabling
Install the switch
Install the network adapter

17 KEEPING THE CUSTOMER INFORMED

Use your answers from exercise 16 to write an email to Robert King informing him about the project.

INFORMATION

Faxe und E-Mails schreiben

Faxe und E-Mails sind meist weniger förmlich als Geschäftsbriefe. Als Begrüßungsformel werden oft die Vornamen benutzt, und die Abschlussformeln sind herzlicher.

Situation	Anrede	Schluss
förmlich	*Dear Ms Belamy,*	*Kind regards,*
neutral	*Dear Peter,*	*Best wishes,*
alltäglich	*Hello/Hi Peter,*	*All the best,*

18 CHOOSING COMPONENTS

a **Match the words in the box to the photos (1–6). There are more words than needed.**

b **Say which of these components you need for setting up a network.**

CAT 5 cable ■ hub ■ modem
network card ■ patchpanel
router ■ switch
telephone socket

1

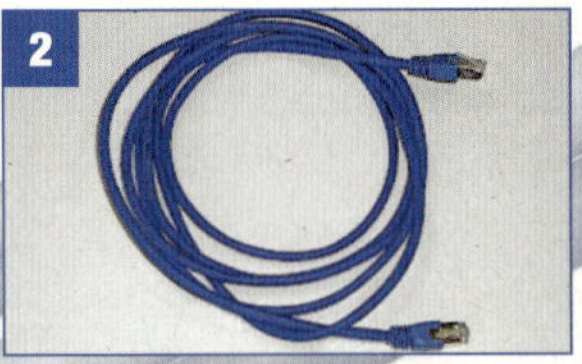
2

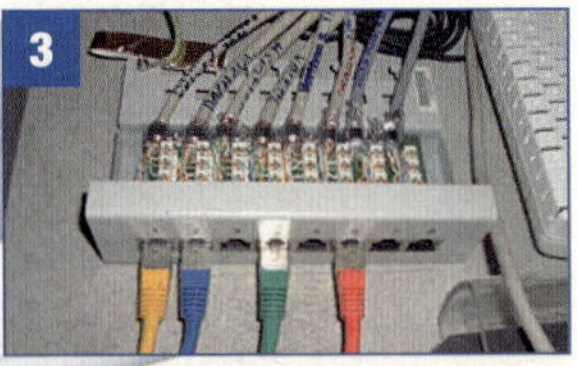
3

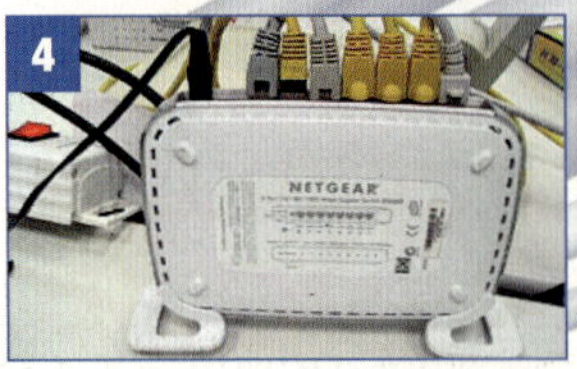

4

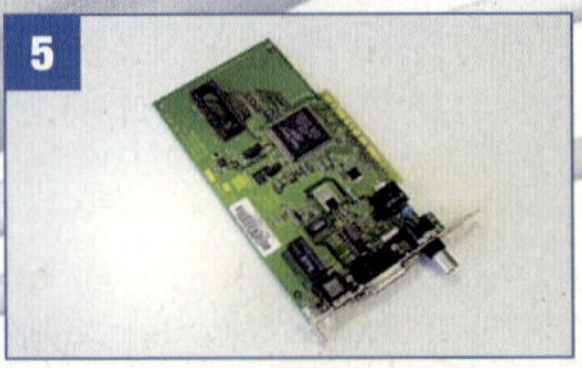
5

6

19 INSTALLING THE SWITCH

Peter Scheller reads the instructions for installing the switch.

check ■ equipment rack
power supply ■ push ■ slot
source ■ tighten ■ ventilation

a Complete the instructions with the words in the box.

b Mediation: Summarize the text in German for a colleague. (The learning box on page 54 will help you.)

INSTRUCTIONS FOR INSTALLING THE SWITCH

1 Assessing the installation site
To operate the switch, the site should have the following facilities:
- Power ...[1] to supply 6.0 A @ 125 V, 60 Hz, or 3.0 A @ 250 V, 50 Hz, for each switch
- Allow at least four inches of space surrounding the switch to provide for proper ...[2].

2 Attaching rack mount brackets
To mount the chassis in an ...[3], attach the rack mount brackets provided.

3 Setting up the chassis

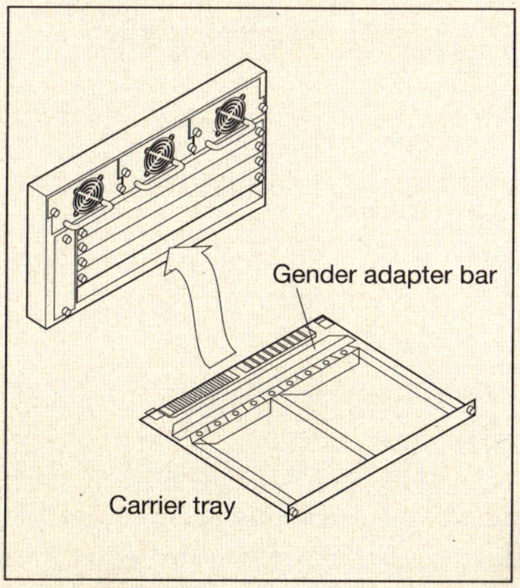

3.1 *Installing the carrier tray*
- Carefully slide the tray into the bottom ...[4] of the chassis.
- ...[5] both capture panel screws
- Carefully slide the module into the left side of the carrier tray.

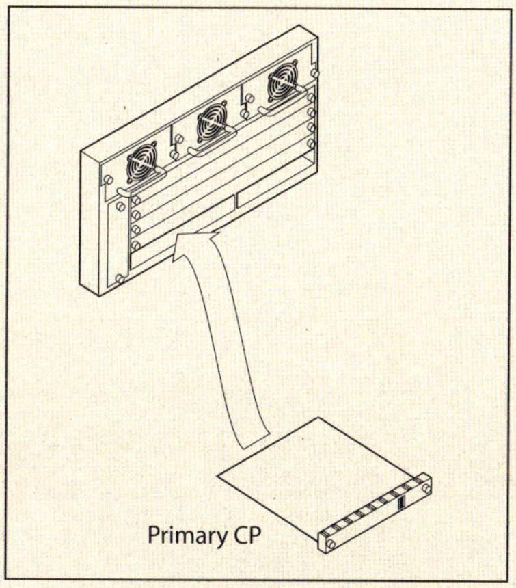

3.2 *Installing the control processor modules*
- ...[6] the module back until it tightly connects with the chassis.

4 Connecting the power cord
- Attach a power cord to each installed ...[7].
- Plug the other end of each power cord into a properly protected AC power source.

5 Checking the physical condition
- Make sure the switch has no physical damage.
- ...[8] the cables for possible crimps or excessive wear that may cause electrical shorts or incomplete connections.
- Make sure the cables are installed according to instructions.

20 INVOICING THE CUSTOMER

Peter Scheller prepares an invoice for King's English.

a **Look at the components: are they similar to the ones you selected in exercise 18? What additional items are there?**

b **Do you think the prices are reasonable? Check them against today's prices on the internet.**

1 Dell Power Edge SC 420 Server	€362,00
5 Dell Dimension 2400	€995,00
1 LevelOne 12 Port 19-Zoll Switch, GSW-1291	€543,44
1 19-Zoll Netzwerk Patchpanel Cat 5, 16 Port	€25,95
1 DSL 4-Port Router	€65,00
6 CAT 5 RJ-45 Steckdosen	€70,00
6 CAT 5 RJ-45 Netzwerkkabel (3 m)	€42,60
100 m CAT 5 Kabel	€89,00
12 CAT 5 RJ-45 Patchkabel	€12,00
1 HP Farb-Laserdrucker 2600 N	€447,00
1 19-Zoll Elmeg-Rack	€325,00
Arbeitskosten: 2 × 5 Stunden à €80,00	€800,00

21 PROJECT: INSTALLING A NETWORK

Work in a small group.

- **Choose a small company and plan a network for all its computers and peripherals.**
- **Present your project to the class.**

LEARNING

Mediation

Manchmal muss man den Inhalt eines Schreibens oder eines Gesprächs an jemanden weitergeben, der die jeweilige Sprache nicht beherrscht. Das nennt man Mediation. Dabei geht es darum, Informationen aus der Muttersprache sinngemäß in die Fremdsprache oder umgekehrt zu übertragen. Dabei genügt es, die wesentlichen Informationen zusammenzufassen, anders als bei einer Übersetzung – dort zählt jeder Satz.

Was sollte beachtet werden?

1 Lesen Sie den Text langsam und sorgfältig.
2 Markieren Sie unbekannte Wörter und schlagen Sie diese im Wörterbuch nach. (Standard-Übersetzungsprogramme sollten Sie meiden!)
3 Markieren Sie die Kernaussagen bzw. Fakten und schreiben Sie diese auf.
4 Versuchen Sie nicht, wörtlich zu übersetzen.
5 Fassen Sie den Text zusammen und bauen Sie die Kernaussagen ein.

File Edit View Go Bookmarks Tools Help

http://www.more.net/technical/netserv/tcpip/firewalls/

Installing a firewall – a step towards network security?

Firewalls provide a barrier that prevents unauthorized traffic between networks.

Selecting the right firewall for your organization can be a time-consuming process. Persuasive advertising with claims of easy installation and management can lead organizations to make hasty decisions.

Organizations should consider the impact a firewall will have on all network services, resources and users, and how a firewall will fit in with their particular business needs and network infrastructure. In some cases, they may find that a firewall solution is unnecessary or impractical.

When implemented correctly, firewalls can control access both to and from a network. They can be configured to keep outside users from gaining access to internal networks and services. They can also be configured to prevent internal users from gaining access to outside networks and services. Many firewalls can be deployed within an organization to restrict access between departments.

On the other hand, firewalls can frustrate users when network resources or services are blocked or are unavailable to them. In addition, a firewall adds to network management responsibilities and often makes network troubleshooting more complex.

The most common misconception about firewalls is that they guarantee security for your network. A firewall cannot guarantee that your network is 100% secure:

- Firewalls cannot offer any protection against inside attacks. A high percentage of security incidents today come from inside the network.
- Firewalls cannot protect against unauthorized access through 'back doors' on your network. Back doors are typically created when an internal user dials out from an unauthorized modem and establishes a connection to an outside network.
- Most firewalls cannot provide protection against viruses or malicious code.

What do we learn from all that? When choosing and implementing a firewall solution, do the homework and make a decision based on your organization's needs and resources.

1 UNDERSTANDING THE TEXT

Read the text and answer the questions.

1 What benefits can a firewall give an organization?
2 What disadvantages can a firewall have?
3 What information does the text give about the level of security a firewall provides?

2 GIVING ADVICE

Use the information from the text to advise Robert King from King's English about installing a firewall for his network.

UNIT 6
Energy and the environment

1 WARM UP

Work with a partner.

a **Name the conventional and alternative forms of energy shown in the picture, using the words in the box.**

biomass ■ fossil fuel ■ geothermal
hydro-electric ■ methane
nuclear ■ solar ■ tidal ■ wind

b **What are the advantages and disadvantages of each form?**

c **Which form(s) of energy would you prefer for powering your house? Why?**

2 READING TRADE LITERATURE

In the magazine *Solar Home* there is an article on solar energy for houses and offices.

a **Before you read the article: discuss the differences between solar thermal collectors and photovoltaic cells with your partner. Make notes in English.**

b **Now read the article. Does it mention the differences that you wrote down?**

Join the solar revolution – power your home with solar energy

The sun's energy is abundant, it's non-polluting and it's free. So why not use it to reduce your energy bill? There are two systems to choose from: 1

SOLAR THERMAL COLLECTORS

Solar thermal collectors can heat domestic water or radiators in your house. They use the energy of the sun, which is on average about 1000 watts per square metre on a clear, dry day on a surface facing the sun. 2

Solar thermal collectors exploit the greenhouse effect by allowing visible and ultraviolet light to pass through a glass plate on top of an insulated collector housing, but trapping the reflected infrared radiation. As a consequence, fluid (usually water or antifreeze solution) circulating in pipes inside the solar collector is heated up and is transferred to an insulated reservoir or used to heat a building directly. 3

PHOTOVOLTAIC CELLS

Photovoltaic (PV) energy generation preserves the earth's finite fossil-fuel resources – coal, oil, natural gas – and reduces air and noise pollution associated with these energy sources. 4

PV systems also protect consumers and businesses from power outages and electricity price increases. Recent technological breakthroughs combined with federal and state programmes and local incentives have greatly reduced the costs of owning a PV system. 5

Depending upon where you live and taking into account the current pricing of public utilities, PV systems can be competitively priced with traditional energy sources. In addition, PV system reliability and durability are outstanding – typical PV systems may last 30 years with minimal maintenance. 6

How do PV cells work? The basic principle is simple. The cells, which are made of silicon, convert sunlight directly into electricity. When light strikes the cell, a certain portion of it is absorbed by the semiconductor material. Electrons are knocked loose by this energy and start flowing. As PV cells have an electric field, the electrons are forced to flow in a certain direction. Contacts on the top and bottom of the PV cells feed the current into the grid. 7

SOLAR ENERGY – A STEP INTO A CLEAN ENERGY FUTURE

Thermal collectors and PV cells may not always be the cheapest way to power your home, but they are definitely the cleanest. Fossil fuels have already been depleted and are running out. Meanwhile environmental pollution is steadily increasing. Solar power is a realistic alternative which can supply our homes with clean energy – now and in the future. 8

3 WORKING WITH WORDS

a Find the English equivalents of these words in the text.

1 Heizkörper (2. Absatz)
2 Flüssigkeit (3. Absatz)
3 Energiequelle (4. Absatz)
4 begrenzt (4. Absatz)
5 öffentliche Versorgungsbetriebe (6. Absatz)
6 Zuverlässigkeit (6. Absatz)
7 Haltbarkeit (6. Absatz)
8 Umweltverschmutzung (8. Absatz)

b Now use the English words to complete the text below.

CLEAN, GREEN AND RELIABLE

We rely on ...[1] to provide power for our homes and businesses. At present, much of this power comes from 'conventional' ...[2] such as fossil fuels. However, these are ...[3]: they won't last forever. They also cause serious ...[4].

We can reduce these problems by using the sun's energy. There are two systems available. Solar thermal collectors contain ...[5] which is heated by the sun. They are used to heat water for ...[6], showers, etc.

PV systems use the sun's energy in a different way, to generate electricity. Their ...[7] is excellent – they rarely malfunction – and their ...[8] is even better: they can last for 30 years or more.

4 MEDIATION

A German customer asks you for advice about the benefits of solar energy systems. Summarize relevant information from the article on page 57. Do not translate word-for-word.

5 ADVISING A CUSTOMER

12

Maren Sørensen, a Danish customer, contacts solar power specialists SunPower to find out about installing a solar system for her house. Christian Götz of SunPower goes to see her.

a Before you look at the dialogue, discuss what you think the customer will want to know.

b Read or listen to the dialogue. Decide whether the following statements are true, false or not in the text. Correct the false statements.

1 Maren wants a solar thermal collector.
2 Her house is ideal for a solar system.
3 A module is the smallest part of a PV system.
4 It rains a lot where Maren lives.
5 The size of the PV system mainly depends on the size of the house.
6 Utility bills help SunPower to calculate the size of PV system needed.
7 Generating renewable energy costs more than increasing energy efficiency.

CHRISTIAN So, have you thought about what kind of solar system you want?

MAREN I have an idea, but can you talk me through the different options?

CHRISTIAN Sure. Well, there are two solar systems which could be installed in your house. You could have solar thermal collectors to produce hot water, or you could have a PV system to produce electricity.

MAREN Sorry, what's PV?

CHRISTIAN It stands for 'photovoltaic'. In other words, producing electricity from light.

MAREN I see ... A PV system would reduce my electricity bill, wouldn't it?

CHRISTIAN That's right.

MAREN That would be great. I would want a PV system, then. But do you think my home is suitable for solar power at all? After all, we don't live in an area with a lot of sunshine.

CHRISTIAN At first sight, your house looks perfect for solar power. The roof faces south, so it has the right orientation, and the roof inclination is good too: sunlight will strike the PV cells at an angle close to 90 degrees. That will enable the cells to absorb enough energy all year round. Also, I can't see any trees or buildings nearby which could shade the PV modules.

MAREN Good. Now, you mentioned 'cells' and 'modules'. What's the difference?

CHRISTIAN Well, a PV cell is the smallest unit of a PV system. The cells are connected electrically to form a PV module – or PV panel as it is sometimes called. A module is about a metre wide and a metre and a half long. Often you'll have several interconnected PV modules. That's called an array.

MAREN I see. So how many PV modules would I need?

CHRISTIAN That depends on the weather in your area and your household demand. We refer to meteorological data which give the average monthly sunlight for your area. They take rainfall, cloudy days and other factors into account. We usually suggest designing the system for the worst month, so that you'll produce enough electricity the whole year round.

MAREN That makes sense. What about my household demand? How do you work that out?

CHRISTIAN We use your utility bills to find out your energy consumption for the last year: how many kilowatt hours of electricity you used each month. That enables us to estimate your demand – how much energy you will need.

MAREN Right. I think we use too much electricity at the moment, but a solar system will help us to be more environmentally friendly.

CHRISTIAN Yes, it certainly will. But you should also look at increasing the energy efficiency of your home, so that you don't waste energy ...

6 WORKING WITH WORDS

Explain these terms from the text on page 59 in your own words.

1	orientation	5	PV array
2	inclination	6	household demand
3	PV cell	7	energy consumption
4	PV module	8	energy efficiency

7 READING NUMBERS

Work with a partner.

Student A: Turn to file 3 on page 127.
Student B: Turn to file 12 on page 130.

LANGUAGE

Numbers and mathematical operations

Numbers

163	'one/a hundred and sixty-three'
3,545	'three thousand five hundred *and* forty-five'
1,000,000	'one/a million'
1,000,000,000	'one/a billion' (German: *Milliarde*)
1,000,000,000,000	'one/a trillion' (German: *Billion*)
12^8	'twelve to the power of eight'
2 m^2	'two square metres'
10 cm^3	'ten cubic centimetres'

Fractions

½	'half'
⅓	'one/a third'
¼	'one/a quarter'
⅙	'one/a sixth'
⅚	'five sixths'

Decimals

In English a comma is used to mark thousands, and a point to show decimals.

2.596	'two point five nine six'
0.954	'nought point nine five four' (BE)
	'zero point nine five four' (AE)

Mathematical operations

addition	+	'plus'
subtraction	–	'minus'
multiplication	× or *	'times' / 'multiplied by'
division	÷ or _ or /	'over' / 'divided by'
	=	'is' / 'is equal to' / 'equals'

Brackets

( ) 'round bracket'
[ ] 'square bracket'
{ } 'curly bracket'

° C = (° F – 32) * 5/9 'Degrees Celsius equals – open bracket – degrees Fahrenheit minus thirty-two – close bracket – times five divided by nine.'

For more numbers and symbols, refer to p. 191 at the back of the book.

8 READING AND COMPLETING A TABLE

Copy the table and complete it for your own home.

Worksheet to determine required number of solar modules for your home

Method	Average home in the USA	Your home
Average monthly kilowatt hours used (from your electricity bill)	600 kilowatt hours	
Multiply by 1,000 for watt hours in a month.	600 × 1,000 = 600,000 watt hours	
Divide by 30 for total watt hours used in an average day.	600,000 watt hours ÷ 30 days = 20,000 watt hours per day	
Subtract watt hours you will save through energy efficiency.	20,000 – 6,400 = 13,600 watt hours per day	
Multiply by the percentage of your home you want to power with the system.	100% × 13,600 = 13,600 watt hours per day	
Divide by maximum (June) insolation (7.5 kWh/m²/day).	13,600 watt hours per day ÷ 7.5 kWh/m²/day = 1,813 watts	
Multiply by 1.2 to cover system inefficiencies.	1,813 × 1.2 = 2,176 watts	
Divide by the peak wattage of the module.	2,176 watts ÷ 120 peak watts per module = 18.13 modules	
Round up to a whole number.	19 modules	

9 DISCUSSION

Read the Culture box, then discuss in a group:

What do you think is the best way of producing clean energy: nuclear energy, alternative sources like wind power, or making fossil fuels cleaner?

CULTURE

Stromerzeugung und Energiepolitik

Die Energiepolitik weist von Land zu Land starke Unterschiede auf: Während z.B. in **Deutschland** ca. 60% des Stroms aus fossilen Energieträgern (Steinkohle, Braunkohle, Öl und Erdgas) erzeugt wird, sind es in **Großbritannien** ca. 90%.

Das wird sich in den nächsten Jahren allerdings ändern müssen, weil die bisher reichen Öl- und Gasvorräte der Briten zur Neige gehen werden. Immer mehr Windkraftanlagen werden dort gebaut, aber auch neue Kernkraftwerke werden geplant.

Weltweit größte Verbraucher und auch Erzeuger von Strom sind die **USA**. Ca. ein Viertel allen Stroms wird in den USA erzeugt und verbraucht. Dort haben fossile Energieträger mit 70% den weitaus größten Anteil an der Stromerzeugung. Kernenergie trägt 20% bei und alternative Energiequellen wie Wind, Photovoltaik, Geothermik oder Biomasse gerade noch zwei Prozent.

10 GRID-TIED OR STAND-ALONE?

13 Christian Götz is helping Maren Sørensen to choose the best PV system for her needs.

a Listen to the dialogue and label the schematics with the expressions in the box. You need some of the expressions more than once.

charge controller ■ inverter ■ meter ■ national grid ■ PV array

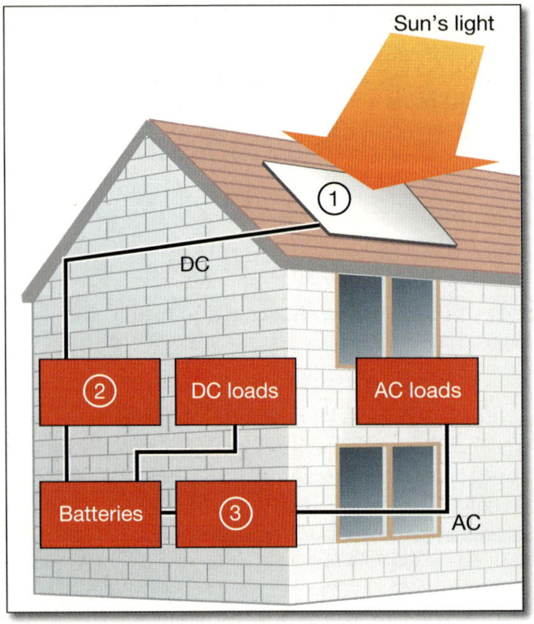

Stand-alone photovoltaic system

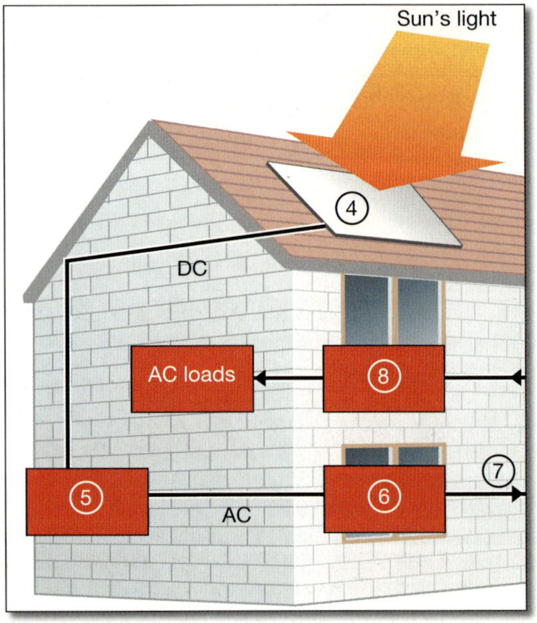

Grid-tied photovoltaic system

b Copy the table into your exercise book. Then listen again and complete the table.

System	Advantages	Components needed
Stand-alone		
Grid-tied		

LEARNING

Wie kann ich Grammatik lernen?

1 Suchen Sie sich eine Grammatik, die zu Ihrem Kenntnisstand passt.
2 Lernen Sie die wichtigsten Begriffe (z.B. Präsens, Perfekt, Adjektiv, Pronomen usw.).
3 Machen Sie sich Ihr grammatikalisches Problem klar.
4 Suchen Sie sich die Informationen dazu im Grammatikbuch (Stichwortverzeichnis).
5 Achten Sie beim Lesen englischer Texte besonders auf diese Form.
6 Um eine Struktur zu lernen: Entwerfen Sie ein Poster mit Beispielen und Regeln, das Sie in Ihrem Zimmer aufhängen und häufig sehen können.

GRAMMAR

The passive

1 It **is absorbed** by the semiconductor material.
2 Electrons **are knocked** loose by this energy.
3 Fossil fuels **have** already **been depleted**.
4 Two solar systems **could be installed** in your house.
5 The meters **will be installed** by us as part of our service.

- Das Passiv wird oft verwendet, wenn man auf eher unpersönliche Art und Weise über Fakten, Vorgänge und Zahlen berichtet.
- Das Passiv wird mit der entsprechenden Form von *be* + Partizip Perfekt (3. Verbform) gebildet. Wie im Aktiv können alle Zeiten gebildet werden.
- Passivsätze mit modalen Hilfsverben werden mit einem modalen Hilfsverb + *be* + Partizip Perfekt (3. Verbform) gebildet. (4)
- „Von" und „durch" wird in Passivsätzen durch *by* ausgedrückt. (1, 2, 5)

11 PRACTICE

Complete the text with the correct passive forms of the verbs in brackets.

Our planet ... (damage)[1] by environmental pollution already and this problem ... (make)[2] even worse by global warming. It's high time something ... (do)[3] before it's too late .

Today renewable forms of energy ... (support)[4] by most European governments. They argue that a lot of oil ... (save)[5] in the future by generating energy with wind turbines or PV systems. Another positive outcome will be that new jobs ... (create)[6] in the field of alternative energy.

As a result of these initiatives, last year more PV arrays ... (put)[7] on house roofs than ever before.

New and more efficient renewable energy technologies ... (invent)[8] every year. In future environmentally friendly fuels like hydrogen ... (use)[9] in cars.

12 SOLAR ENERGY FAQS

Complete the answers to these FAQs by using the passive forms of the verbs. Use the same tense as in the question.

PV FAQs

1 **How can I reduce my electricity bill?**
It … (can / reduce) by installing a solar energy system in your home.

2 **Do solar thermal collectors generate electricity?**
No, electricity … (not / generate) by solar thermal collectors. You need a PV system.

3 **If I install a PV system, can I sell electricity back to the grid?**
Yes, electricity … (can / sell) to the grid.

4 **Will I need two electricity meters for my PV system?**
Yes, two meters … (need) for a grid-tied system.

5 **Can I maintain the system myself?**
No, the system … (can / not / maintain) by the owner. Expert maintenance is needed.

6 **Will PV systems help to reduce CO_2 emissions?**
Yes, CO_2 emissions … (reduce) by the use of PV and other renewable energy technologies.

13 PV INSTALLATION

Work with a partner.

Student A: Turn to file 4 on page 127.
Student B: Turn to file 13 on page 130.

LANGUAGE

Describing a process

First(ly) / first of all (the PV array is installed).
The first step is to (install the PV array).
Next / After that / Then (the charge controller is set up).
The last step is to (connect the system to the grid).
Finally (the system is connected to the grid).

14 OPERATING MANUAL FOR AN INVERTER

a **Complete the text using the words in the box.**

b **Summarize important information about the inverter for a German customer. Do not translate the text word-for-word.**

moisture ■ emergency stop button
conditions ■ generating
Sleep Mode ■ shutdown
maintenance ■ personnel

PV-1280

SAFETY FEATURES

Warning
The PV-1280 enclosure contains exposed high voltage conductors. The enclosure door should remain closed, except during …[1] or testing.
The servicing instructions are for use by qualified …[2] only.
To reduce the risk of electric shock, do not perform any servicing other than that specified in the operating instructions unless you are qualified to do so. Do not open the cabinet door if extreme …[3] is present (rain or heavy dew).

Front Panel Indicators

The PV-1280 incorporates three coloured LED indicators, used to show the current operating state of the inverter. The indicators have the following meanings:

Red Fault Mode. The inverter has sensed an abnormal condition. To reset the unit (clearing the fault condition), return the …[4] to its extended position (see below*).

Yellow …[5]. The inverter is waiting for sufficient PV voltage to start the inverter.

Green Operator Mode. The inverter is active and …[6] alternating current.

Emergency Stop Button

The PV-1280 incorporates a red emergency stop button (E-Stop) located on the front door of the enclosure. Under normal …[7], the E-Stop is in the extended (out) position. Depressing the E-Stop will initiate a controlled …[8] of the PV-1280 and open the main contactor within the unit. The PV-1280 is prevented from being restarted until the E-Stop is returned to the extended position.

Do the energy efficiency quiz.

ENERGY EFFICIENCY – How do you measure up?

1 You decide that you need a two-week holiday. Which is the most eco-friendly choice?
- **a** A winter sports holiday, skiing in the Italian Alps
- **b** A yoga retreat in India, with meditation and simple living
- **c** A driving holiday in your own country

2 You decide to go to India, but you know the flight will emit some CO_2 into the atmosphere and you want to offset it. How many trees will you have to plant so that the CO_2 is absorbed in 100 years?

a 0.2 **b** 2 **c** 20

3 Your clothes are dirty. Which is the most eco-friendly way to clean them?
- **a** Hand wash them in hot water.
- **b** Take them to the dry cleaners.
- **c** Machine wash them in cold water.

4 Your father wants to redesign the garden. Which choice of planting is the least eco-friendly?

a Grass **b** Shrubs **c** Trees

5 You have a new baby sister and your parents have to decide what kind of nappies they are going to use. Which of these is the best choice for the environment?
- **a** Use disposable nappies.
- **b** Use a laundry service which picks up dirty nappies and returns them clean.
- **c** Wash and dry the baby's nappies at home.

6 You are thirsty, so you buy a takeaway drink. Which is the most eco-friendly choice?
- **a** A plastic bottle of mineral water bottled at the source in a French town
- **b** A can of Coke™
- **c** A hot coffee in a polystyrene cup

7 You are tidying up your house, going back and forth between the bedroom, kitchen and living room, spending five to ten minutes in each. What is the best way to make sure your lights don't waste electricity?
- **a** Keep the lights on as you go from room to room until the job is done.
- **b** Turn the lights off when you leave a room and on again when you return.
- **c** Turn the lights off when you leave a small room and on when you return, but leave the lights on in large rooms.

8 After all that tidying, you are hungry. You decide to have a tin of soup. What is the most energy-efficient way of heating it up?
- **a** A hotplate and pan of the same diameter
- **b** A pan of a bigger diameter than the hotplate
- **c** A hotplate of bigger diameter than the pan

9 You want to make a really significant contribution to the reduction of CO_2 emissions. Which of these would be of the most benefit over the course of a year?
- **a** Take the train instead of driving a car.
- **b** Hang your washing out to dry rather than using the tumble dryer.
- **c** Work from home one day a week.

(Adapted from: http://www.bbc.co.uk)

UNIT 7
Electrical components

1 WARM UP

Work with a partner.

a Identify the photos of electrical components (1–10). All the names are in the box.

capacitor ■ diode
full-wave bridge rectifier
glass fuse ■ integrated voltage stabilizer
mains input socket ■ resistor
transformer ■ transistor ■ valve

1
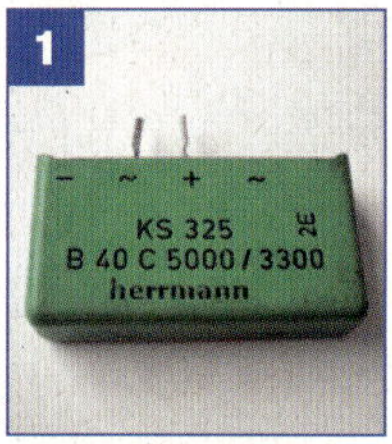

2
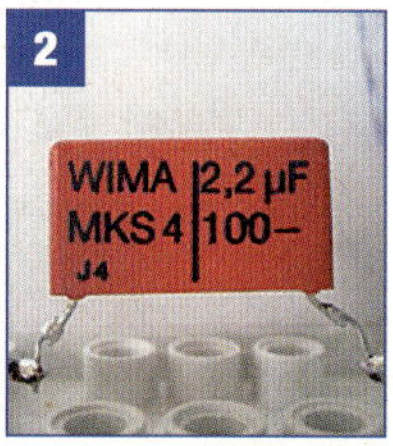

3
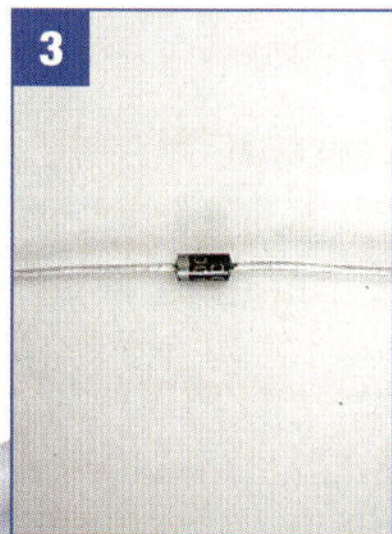

4

5
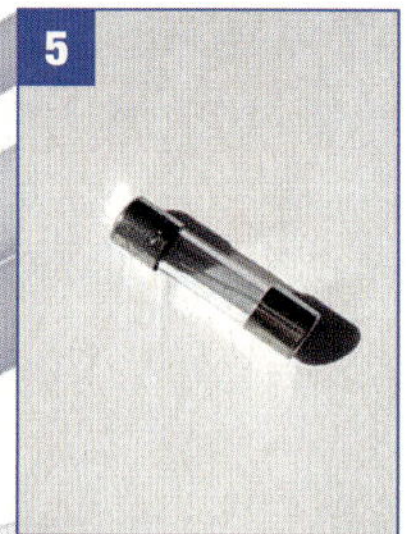

6

7

8

9
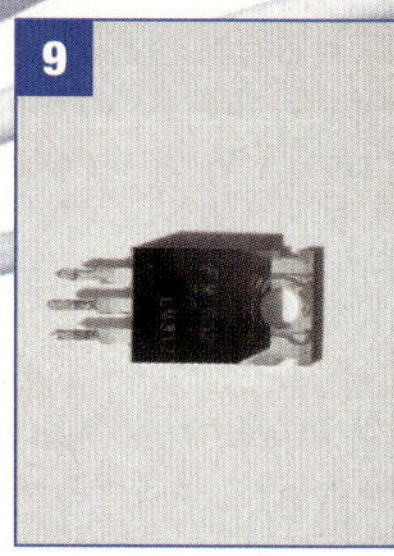

10
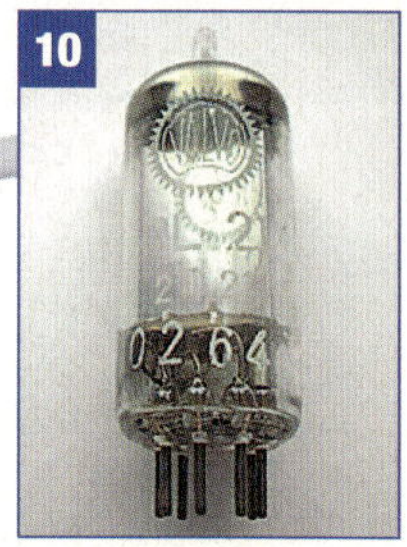

b Find the correct symbol (a–j) for each component.

a
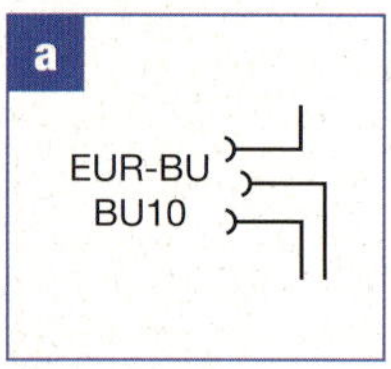

b
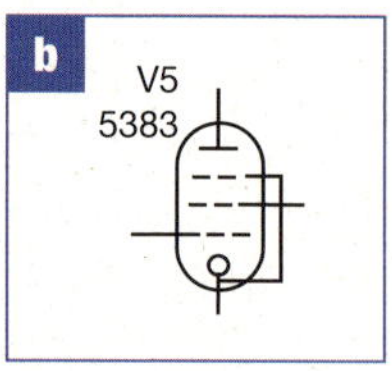

c
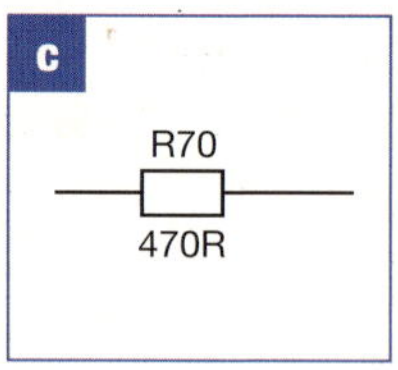

d
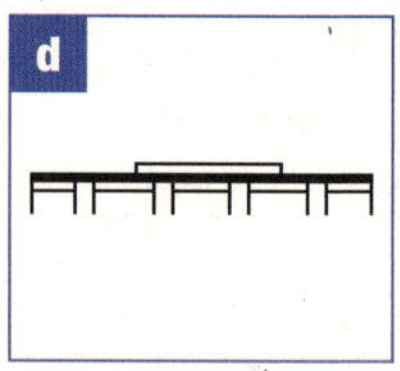

e
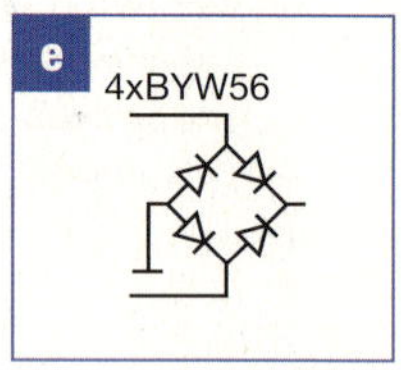

f
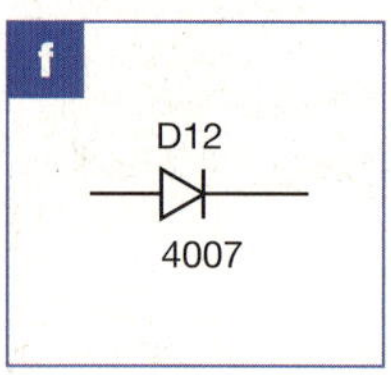

g
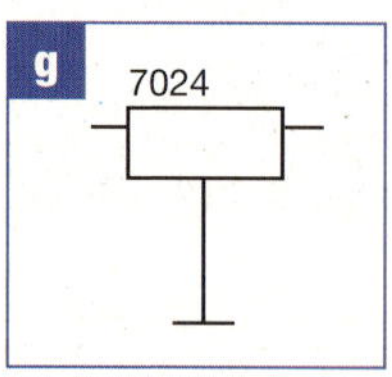

h
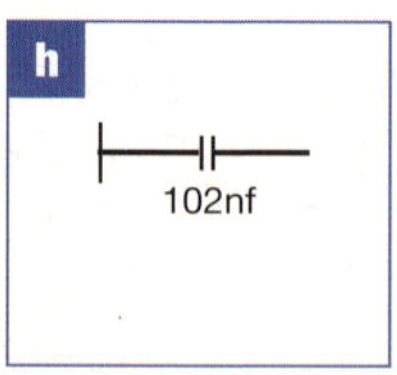

i
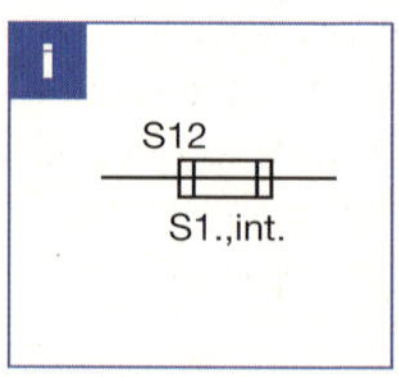

j
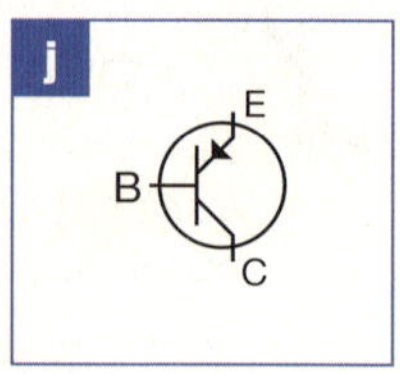

c Give an example of where you would expect to find each component.

2 UNDERSTANDING THE TEXT

Antenna is a monthly magazine for ham radio operators. In this month's issue there is an article by Armin Feiertag, a radio enthusiast from Freiburg im Breisgau.

Read the article and answer the questions below.

1 Who is Tom Vincent?
2 What was the first problem that Armin found?
3 Explain what Armin means when he says 'I carried out a sensory check.'
4 What useful information did Armin find, and where?
5 How did Tom thank Armin for his help?

AM I NOW FAMOUS?

FEATURE ARTICLE BY ARMIN FEIERTAG, KS9OM

LAST SUNDAY MORNING I was having a drink at the Jazzhaus in Freiburg, when I saw Tom Vincent and his band Direct Current arrive to do a sound check for the show that evening. Tom switched on his amplifier but it was dead! He was really worried, so I offered to help. I first checked the fuse and saw that it was blown, so I replaced it. The mains and standby control lights came on but the amplifier still didn't work.

I looked at the back of the amplifier and checked the valves – they were illuminated. I disconnected the AC mains lead at the input socket and removed the back panel of the amplifier. Next I carried out a sensory check and found there was no smell of burning and there were no charred wires. I could see nothing wrong with the amplifier.

Luckily, the circuit diagram of the amplifier was inside the case under the speaker. It showed that the amplifier had a transformer with five output voltages:

- three for the semi-conductors in the amplifier
- one to heat the valves
- one 430 V DC output to power the valves

When I saw this I realized I had to be careful when switching on the amplifier, so I took it back to my radio shack and fixed it there. I got free tickets to the show. Am I now famous?

Armin Feiertag, KS9OM

3 HARDWARE CHECK

Read the text again. Which of these components are not mentioned?

1
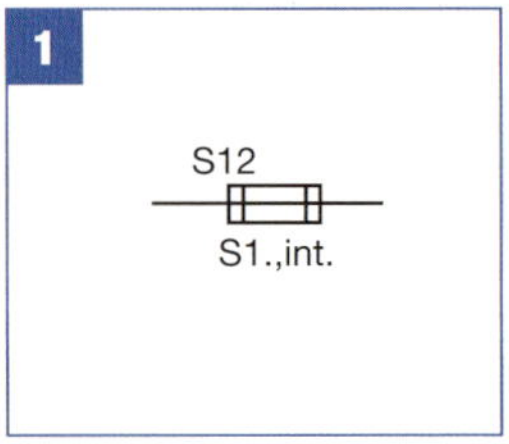

2
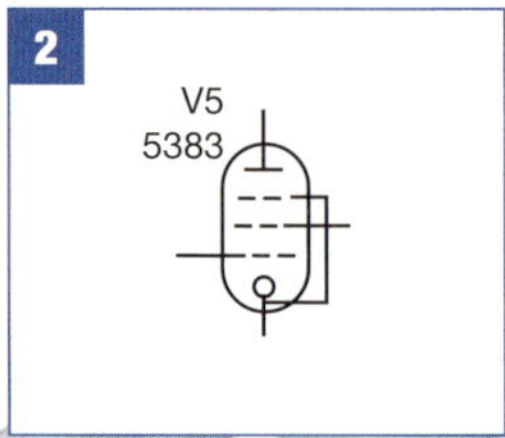

3
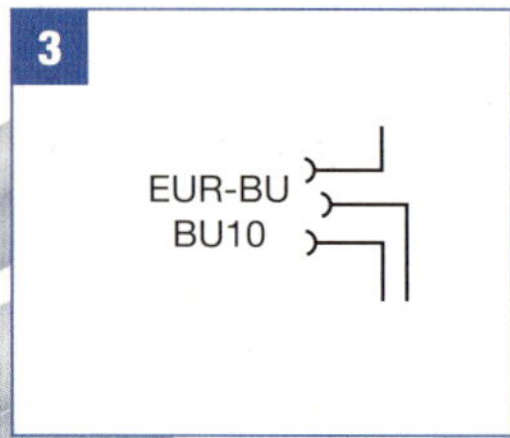

4
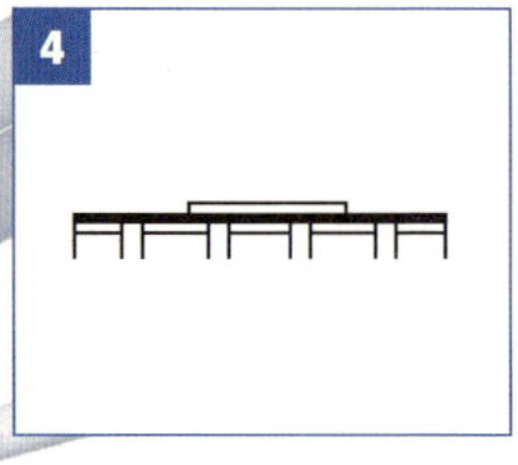

5
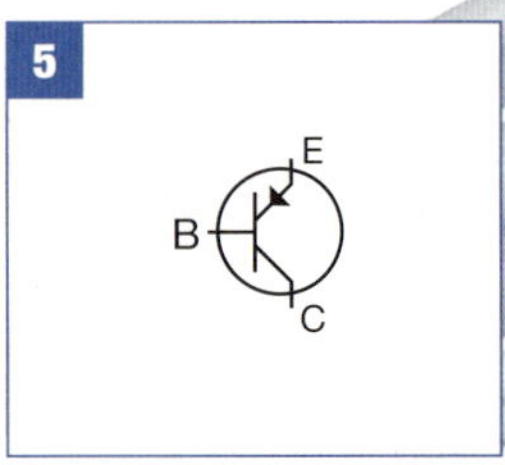

6
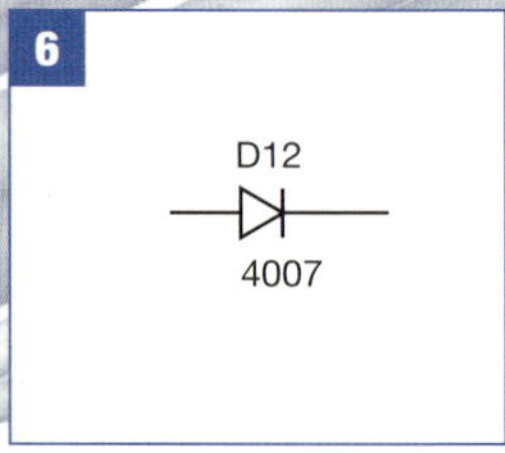

7
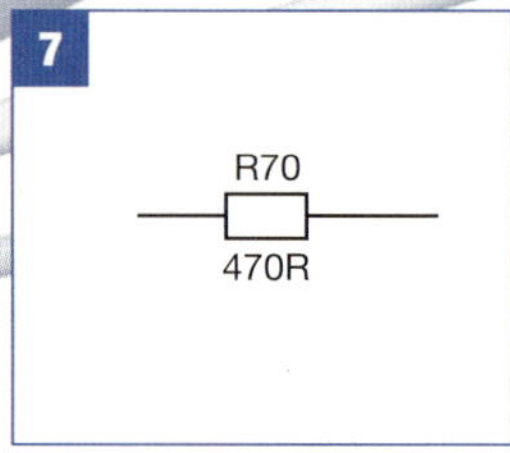

8
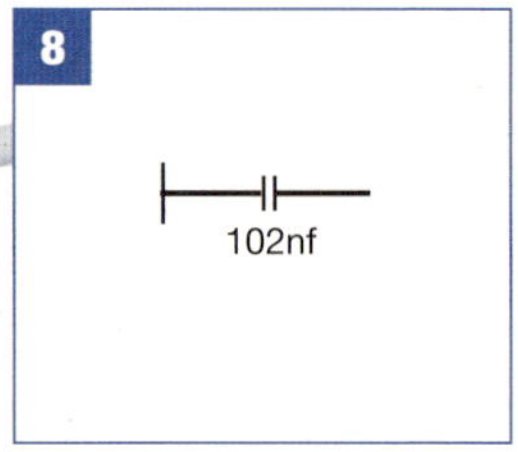

4 DISCUSSION: TRACING THE FAULT

Discuss this question in a small group.

According to the circuit diagram the amplifier has a 430 V DC power supply when on standby, so Armin has to be careful. What would you do next, if you were Armin?

Example: *First I'd make sure that the amplifier was disconnected.*
I'd plug it into a circuit with …
I'd use a … to check the …

LEARNING

Wie man Fachvokabeln mit Hilfe des Internets übersetzt

Wenn Sie eine Fachvokabel im zweisprachigen Wörterbuch nicht finden, können Sie natürlich in einem Online-Wörterbuch wie www.leo.org danach suchen.
Werden Sie dort nicht fündig, sollten Sie nicht gleich aufgeben, sondern es mit ein paar Kniffen versuchen, z.B.:

- Einen deutschsprachigen Wikipedia-Artikel aufrufen und auf die englische Version umschalten.
- Bei Google das Wort durch Eingabe von Kontextwörtern (d.h. Wörtern, die normalerweise zusammen mit dem gesuchten Wort auftreten) suchen. Sie können auch versuchen, das englische Wort zusammen mit deutschen Kontextwörtern zu suchen.
- Technische Seiten wie www.howstuffworks.com aufrufen, den Artikel für das Oberthema suchen und die englisch bezeichneten Teile anhand der Abbildungen erkennen.

Sie möchten z.B. wissen, wie „Entlötlitze" auf Englisch heißt, aber Ihr Wörterbuch enthält dieses Stichwort nicht. Sie rufen die deutschen Seiten von Wikipedia auf und finden dort einen Artikel über Löten. Dann wechseln Sie zur entsprechenden englischen Seite und erfahren, dass Entlöten *desoldering* heißt. *Desoldering wicks* werden in diesem Zusammenhang erwähnt: Ist das vielleicht der Ausdruck, den Sie suchen? Eine schnelle Google-Suche nach „desoldering wick entlötlitze" bestätigt dies.

5 IDENTIFYING COMPONENTS

Work with a partner. Look at the circuit diagram (schematic) below and take turns to ask and answer questions about it. Use expressions from the Language box.

Ask at least five questions each.

Example: *A* *What's the component in square E6?*

B *It's a rectifier.*

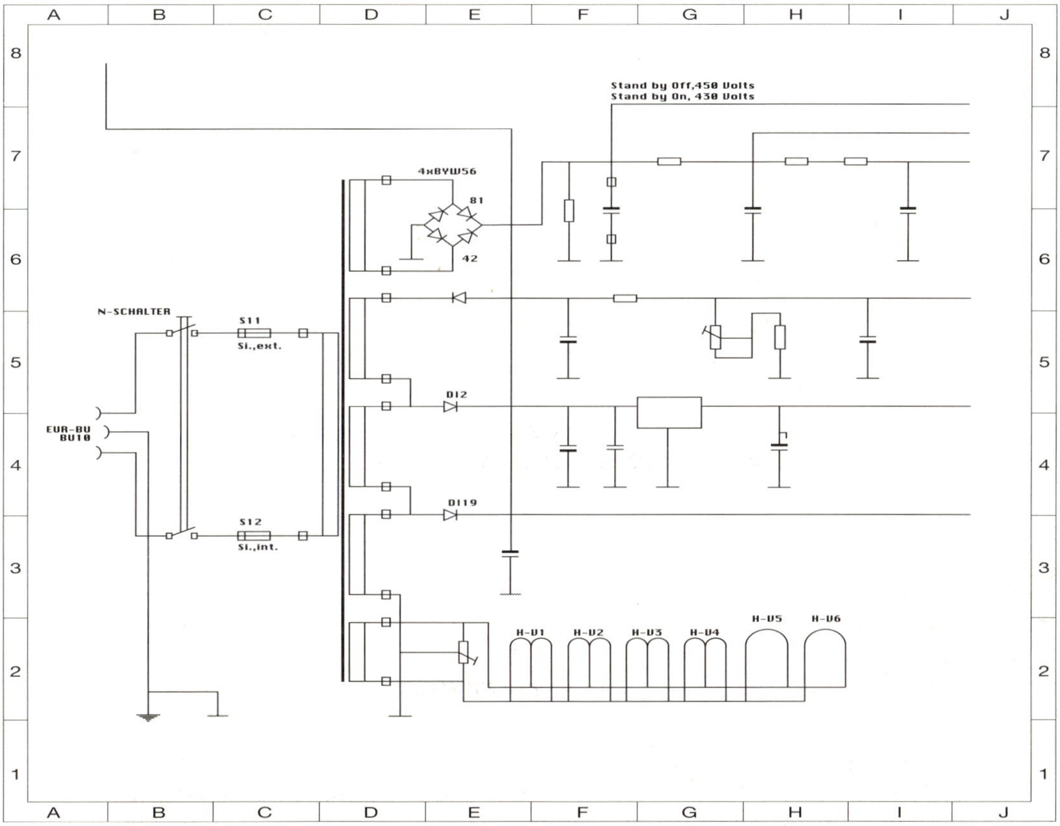

LANGUAGE

Describing circuits

What's the component in square … ?
What's it attached to?
What is there between … and … ?
How many … are there in the circuit?
Where is it / are they?

6 GETTING A SECOND OPINION

14

Armin wants a second opinion on the guitar amplifier, so he contacts Carla, an American ham radio operator.

Listen to their conversation and complete the sentences below.

1 Carla's call sign is …
2 Armin tells Carla that he has replaced …
3 Armin's sensory check showed that there was no …
4 Carla hopes Armin's shack has …
5 Armin traced the signal using a …
6 The voltage is zero where it should be …
7 The transformer is attached to a …
8 Carla tells Armin not to …
9 Carla tells Armin to measure …
10 When it's midday in Germany the time in Portland Oregon is …

7 TRACING THE FAULT, 2

a Match 1–4 to a–d to find possible causes of the fault in the guitar amplifier.

1 A small metal object (e.g. a paper clip) has fallen into the amplifier …
2 There is a dry solder joint …
3 One of the wires to the circuit board was badly soldered …
4 The transformer is faulty …

a … so it came loose and short-circuited the diodes.
b … and is producing no voltage.
c … because the soldering iron used was at the wrong temperature.
d … and has caused a short circuit.

b Decide which is the most probable cause.

CULTURE

Ham radio

Funkamateure werden auf Englisch *radio hams* genannt. Der Amateurfunk ist sehr vielfältig: Es gibt auf der einen Seite diejenigen, denen es in erster Linie auf die Funkverbindung und das Gespräch mit anderen Funkamateuren auf der ganzen Welt ankommt – das nennt man *DXing*. Auf der anderen Seite gibt es die Techniker, die ihre Funkanlage selbst bauen und dann ausprobieren wollen. Bei Krisen wie Naturkatastrophen und Terroranschlägen kann der Amateurfunk einen wichtigen Beitrag zur Koordinierung der Rettungsdienste leisten.

Dem Amateurfunkdienst sind diverse Frequenzbereiche, die sogenannten *amateur bands*, zugewiesen: zwischen 135 kHz und 250 GHz im Langwellen-, Mittelwellen-, Kurz- und Ultrakurzwellen- bis hinauf in den Gigahertzbereich.

Die Amateurfunkstelle wird umgangssprachlich *radio shack* genannt. Oft handelt es sich um ein zweckentfremdetes Schlafzimmer, einen Gartenschuppen oder eine Garage.

Man sollte den Amateurfunk nicht mit dem sogenannten *citizens' band radio (CB radio)* verwechseln, dem ein Frequenzband um 27 MHz zugewiesen ist, und der eine sehr begrenzte Reichweite (20–80 km) hat.

Adapted from http://de.wikipedia.org/wiki/Amateurfunk

8 AN ELECTRONICS FORUM

Armin is browsing through the forum on the 'Electronics online' website when he finds a question about amplifiers from 'Don'. Thinking about his recent experience with the amplifier, Armin posts a reply.

Look at the forum page. The last 4 paragraphs (A–D) of Armin's reply have become mixed up. Put them in the right order.

http://forum.electronicsonline.com/

Electronics online

The internet forum for electronics technology

http://forum.electronicsonline.com

Topic: Dead amplifier?

Posted by Don on 12/21/20.. at 15.35

Hi there!

Can anyone help me with some info on diodes and rectifiers? The full-wave bridge rectifier in my amplifier has an output of 0 volts! Does this mean I can throw away the amplifier now? Any help would be much appreciated.

Cheers!

Posted by Armin on 12/22/20.. at 09.20

Hi Don.

I had a problem like that with an amplifier. Here's a link to the schematic so you can see what type of amplifier it was.

A Now I knew I had to replace them, so I looked them up on the schematic and saw I needed four BYW 56 diodes. I found them on the internet and saw that a BYW 56 diode had a maximum forward current of 2 A and a maximum reverse voltage of 1,000 V. Using my ham radio contacts I found someone with suitable diodes nearby. I got the diodes, soldered them in, connected the wires of the power supply to the circuit board, plugged in the mains lead and checked the DC voltage again. It was 430 V!

B I opened up the amplifier, and carried out a sensory check: no obvious problems. So I plugged it in and used a DMM to check the voltage at the mains input and the rectifier. It was 230 V at the mains input but zero at the rectifier. This showed me that the fault was somewhere between the mains input and the smoothing circuit, so I took a closer look at the rectifier. I then noticed that the diodes were discoloured and one of the wires leading from the power supply to the circuit board had become loose and was touching the metal base, leaving a small black mark.

C Next I needed to check all four diodes but to do this I had to take them out. After disconnecting the AC mains lead, I unsoldered them carefully using a soldering iron and a desoldering pump to suck off the solder and tested them one by one with the multimeter. I set the multimeter to resistance mode and discovered that the diodes showed a high resistance in both directions and were consequently defective.

D I disconnected the mains supply and unsoldered the transformer outlet to the rectifier. Then I reconnected the mains supply and checked the output of the transformer through this outlet. It was 308.5 V AC as expected.

9 WORKING WITH WORDS

Complete the text using the words in the box. (They are all in the forum on page 71.)

amplifier ■ circuit board
desoldering pump ■ diodes ■ disconnect
mains ■ replace ■ resistance
sensory check ■ voltage

ELECTRICAL HANDBOOK

Many electronic devices have an …[1] to make the sound louder. Some are battery-operated and others are plugged into the …[2].

If a device breaks down, it's best to …[3] it first (otherwise you could get an electric shock). Then you should carry out a simple …[4], before taking it apart and replacing electronic components like …[5]. Service personnel normally use a digital multimeter to measure the …[6] (V) or the …[7] (Ω) at test points inside the equipment. Sometimes the components are mounted on a printed …[8] and when a component has been removed, the contacts must often be cleaned with a …[9]. In most cases it's best to …[10] a faulty component.

10 TOM VINCENT'S RADIO INTERVIEW

15

Tom has just finished his set at the Jazzhaus and is talking to a local radio reporter.

Read or listen to the dialogue, then complete the exercise on the next page.

REPORTER You were great, Tom! Everybody loved it!

TOM Thank you! I just love Germany and everybody here in Freiburg.

REPORTER Have you ever been to Freiburg before, Tom?

TOM No, I haven't but I've been to Germany a few times. I came over two years ago when I brought out my first album Electricity.

REPORTER How many copies have you sold of that album so far, Tom?

TOM About a million but I hope to sell more of my new album Resistance. It came out last week and since then it has sold over 100,000 copies.

REPORTER That's great, Tom! Is there a track from your new unplugged album you'd like us to play for our listeners?

TOM Well, I love them all but why don't you play 'Sock it to me, baby!'? It's a cool song with cool lyrics.

REPORTER OK, we'll do that, because our German listeners haven't heard it yet.

TOM Just one more thing before you play my song. I'd like to thank the guy who helped me before the show – I think his name is Armin. My favourite amp didn't work when I arrived here, so he fixed it in his radio shack. I want to dedicate this song to him.

REPORTER That's great, Tom. This one's for you, Armin!

a The reporter wrote an article about Tom Vincent but she made some mistakes. Compare her article with the dialogue and find the mistakes.

Example: *Jazzclub – Jazzhaus*

SOCK IT TO US, TOM!

TOM VINCENT played at the Jazzclub in Freiburg last night. He has never been to Germany before but he loves it. Tom brought out his first album called *Electricity* last year and his latest one came out last month. Since then it has sold 500,000 copies! It's called *Sock it to me, baby!* After the show, Tom thanked his good friend Armin, who fixed Tom's guitar in his radio shack before the concert.

b Write your correct version of the article.

Example: *Tom Vincent played at the Jazzhaus …*

GRAMMAR

The present perfect and simple past

1 How many copies **have** you **sold** of that album <u>so far</u>?
2 <u>Since then</u> it **has sold** over 100,000 copies.
3 Our German listeners **haven't heard** it <u>yet</u>.
4 I **came** over <u>two years ago</u>.
5 … the guy who **helped** me <u>before the show</u>.
6 My favourite amp **didn't work** <u>when I arrived here</u>.

Present perfect

- Mit dieser Zeitform sagt man, dass (nicht: wann!) etwas geschehen ist. Man drückt damit auch aus, seit wann oder wie lange ein Zustand schon andauert.
- Sie wird mit *have/has* und der 3. Form des Verbs gebildet.

Simple past

- Diese Zeitform verwendet man, um zu sagen, wann etwas geschehen ist, oder um über Vergangenes zu berichten. (4–6)
- Sie wird mit der 2. Form des Verbs gebildet (siehe Unit 2, S. 22). (4–6)
- Bestimmte Zeitangaben verlangen gezielt und ausschließlich nach dem *Present perfect* bzw. dem *Simple past*.
  Present perfect – Zeitadverbien wie for *two years, this week, since 1998, already* oder *yet*, da der Zeitraum, auf den sich der Satz bezieht, bis in die Gegenwart reicht.
  Simple past – Zeitadverbien wie *two years ago, last week, When … ?* oder *in 1998*, die sich nur auf die Vergangenheit beziehen.

11 ONLINE CHAT WITH TOM VINCENT

Tom is answering questions from his fans. Complete the sentences with the present perfect or the simple past.

http://www.tomvincent.info/chatroom.php

Chat with Tom

Sabrina from Freiburg
Hi Tom! I love your music! … (you/sell)[1] all the tickets to your next show in Berlin yet?

Tom
I don't know, Sabrina. My manager … (not tell)[2] me yesterday when I … (see)[3] him in the recording studio.

Sabrina
Oh dear! I hope I get a ticket!

Sascha from Hamburg
Hi Tom! I love your guitar sound! … (you/ buy)[4] your guitar in Nashville last year?

Tom
I … (buy)[5] three guitars in Nashville last year and since then I … (buy)[6] another one – a 'Tom Vincent single coil special'.

Silke from Wattenscheid
So far I … (see)[7] every one of your concerts! I … (never miss)[8] a chance to see you perform live! Please send me one of your T-shirts, Tom!

Tom
It's on its way, Silke! Take care!

Mareike from Berlin
I just want to say your band is great! When … (you/release)[9] your first record – in 2004 or 2005?

Tom
My first record … (come)[10] out in 2005.

12 IT'S YOUR TURN

Describe an electrical repair you have carried out. Use the verbs in the box to help you.

Example: *Once I had to repair a standard lamp. First, I disconnected it from the mains, then I …*

check ■ (dis)connect ■ examine ■ order a new … ■ plug in ■ replace
screw up ■ shorten ■ test ■ unplug ■ unscrew

MENU»

Know it All

Why are tubes* still used?

High-power RF applications

Many radio stations still use big power tubes, especially for power levels above 10,000 watts and for frequencies above 50 MHz. High-power UHF TV stations and large FM broadcast stations are almost exclusively powered by tubes. The reason is cost and efficiency – only at low frequencies are transistors more efficient and less expensive than tubes.

Making a big solid-state transmitter requires wiring hundreds or thousands of power transistors. They also require large heat-sinks to keep them cool. An equivalent tube transmitter can use only one tube and can be cooled with air or water, thus making it smaller than the solid-state transmitter. The low cost of RF power generated by tubes has kept them economic.

Guitar amplifiers

Worldwide all-tube or hybrid tube-plus-transistor models make up 80% of the market for guitar amplifiers, which is thought to be worth at least $100 million.

Why tube amplifiers? It's the tone that musicians want. The amplifier and speaker become part of the musical instrument. The distortion and speaker characteristics of a tube amplifier, with an output transformer to match the speaker load, is unique and difficult to simulate with solid-state devices.

Even the wildest rock musicians seem to be very conservative about the equipment they use to make their music, and insist on vacuum tube technology.

*(vacuum) tube (US) = valve (GB)

1 UNDERSTANDING THE TEXT

Explain the advantages of valves over transistors (a) when used by radio stations, (b) when used in guitar amplifiers.

2 ABBREVIATIONS

Match the abbreviations in the text with the following definitions.

1 This is the number of million oscillations of electromagnetic waves per second.
2 This is the radio frequency range from 30 MHz (wavelength 10 m) to 300 MHz (wavelength 1 m).
3 This is the radio frequency range from 300 MHz (wavelength 1 m) to 3.0 GHz (wavelength 10 cm).
4 This means varying the frequency of a signal to communicate information.
5 This is the lengths of electromagnetic waves which are used for radio communication and generated by alternating current.

3 MEDIATION

You visit a German-language internet forum about electronics. Someone has posted a question about valves in guitar amplifiers. Summarize the text in German for your reply (approx. 100 words).

UNIT 8
Measuring and testing, 1

1 WARM UP

a Match the names with the instruments.

b Explain what they are used for.

c Choose one of the instruments that you have used recently. What values did you measure and why?

ammeter ■ capacitance meter
mains tester
multimeter ■ ohmmeter
oscilloscope
test light ■ voltmeter

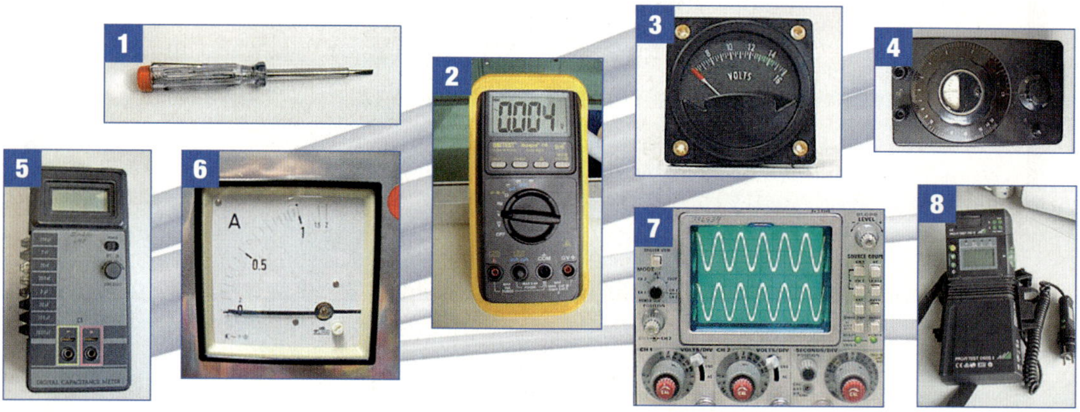

2 UNITS OF MEASUREMENT

Work with a partner.

Student A: Turn to file 5 on page 127.
Student B: Turn to file 14 on page 131.

3 USING A DIGITAL MULTIMETER (DMM)

Look at the DMM on the next page, then decide whether these statements are true or false. Correct the false statements.

1 This DMM can measure both alternating and direct current.
2 It can measure an AC voltage of up to 1,000 volts.
3 It can measure a DC voltage of up to 1,000 volts.
4 It can measure a resistance of 2,000,000 ohms.
5 It can measure a very strong current of up to 2,000 amps.
6 In AC mode it shows the maximum voltage.

4 VOLTAGES

Read the Language box, then work with your partner.

Student A: Turn to file 6 on page 128.
Student B: Turn to file 15 on page 131.

LANGUAGE

Talking about voltages

$$\text{rms} = \frac{\text{Peak voltage (PV)}}{\sqrt{2}}$$

The root mean square (rms) voltage equals the peak voltage over (divided by) the square root of two.

$$\frac{17\text{ V}}{\sqrt{2}} = ?$$

What's the root mean square value of 17 volts?

$$\frac{10{,}000\text{ V}}{1.414} = 7{,}072.136\text{ V}$$

If the peak voltage is ten thousand volts, the root mean square voltage is ten thousand over (divided by) one point four one four, which equals seven thousand and seventy-two point one three five volts.

5 HOW AN OSCILLOSCOPE WORKS

Match 1–5 to a–e to explain important features of an oscilloscope.

1 An oscilloscope draws a graph of
2 The screen has eight squares on the vertical (Y) axis and
3 It displays voltage and time
4 The waveform shows how the voltage
5 An oscilloscope gives an exact picture of a voltage and is

a alternates between positive and negative values.
b an excellent tool for testing circuits and finding faults.
c voltage against time – a V/t graph.
d in the form of a wave.
e ten squares on the horizontal (X) axis.

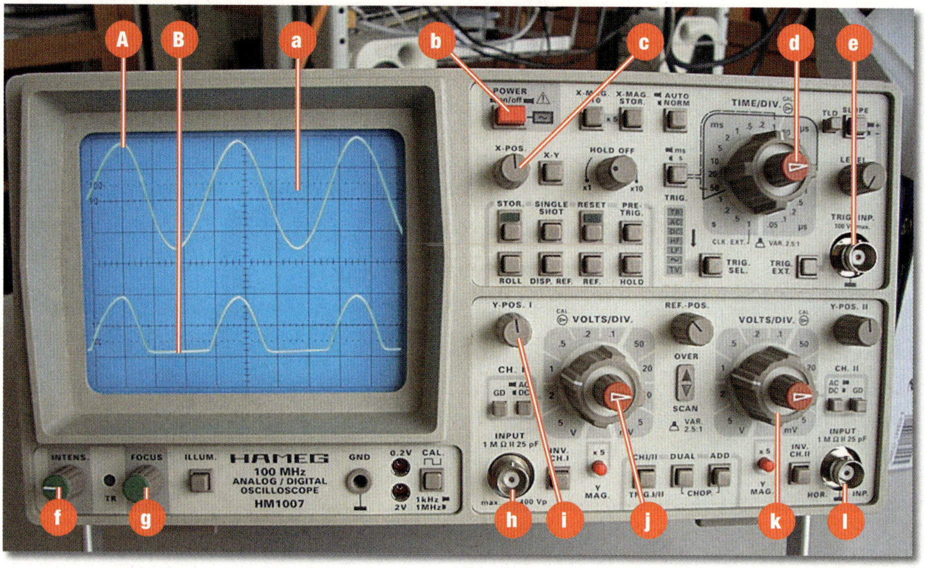

6 OSCILLOSCOPE CONTROLS

a **Look at the oscilloscope and find the control (a–l) which matches each description (1–10).**

1 The three input sockets
2 The time per division control
3 The knob which adjusts the intensity (brightness) of the trace
4 The knob which focuses the trace
5 The knob which moves the waveform to the left and the right
6 The knob which moves the waveform up and down for voltage no. 1
7 The voltage per division control for voltage no. 1
8 The voltage per division control for voltage no. 2
9 The screen
10 The on/off button

b **Choose three of the controls and explain how they work.**

c **Which of the graphs on the screen of the oscilloscope is (1) a sinusoidal waveform, (2) a rectified waveform?**

d **How many waveforms can this oscilloscope show at the same time?**

7 MAKING A BATTERY CHARGER

16

Today the apprentices at Elektro Schillinger have to make a battery charger using a circuit diagram. Dennis Lohmann is working with Matti Junkkari, an exchange student from Finland.

Listen to the dialogue, then answer the questions.

1 List the four stages of the task, as explained by Dennis.
2 Where will the apprentices get the components from?
3 What sorts of batteries will the device be able to charge?
4 What are the apprentices going to search for on the internet?

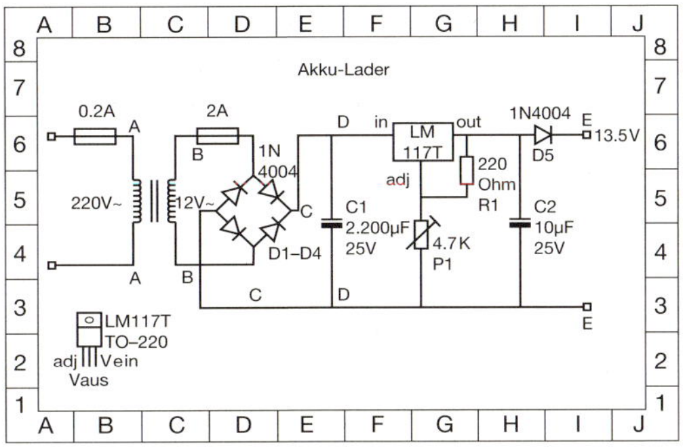

8 DOUBLE-CHECKING

Role-play the dialogue with a partner. Use expressions from the Language box.

Trainee / **English colleague**

1. Trainee: Sie bitten Ihren Kollegen, das Bauteil in Feld C5 zu erklären.
2. English colleague: Es ist ein Transformator.
3. Trainee: Bedanken Sie sich. Fragen Sie, ob Sie mehrere Akku-Lader bauen sollen.
4. English colleague: Sie verneinen und erklären, dass Ihr Chef möchte, dass Sie den Akku-Lader mit einem Oszilloskop testen.
5. Trainee: Sie sagen, dass Sie nicht ganz sicher sind, wo Sie den Akku-Lader testen sollen.
6. English colleague: Sie sollen ihn an den Stellen B, C und E testen.
7. Trainee: Sie fragen, was das Bauteil in Feld D5 ist.
8. English colleague: Es ist ein Zweipulsbrücken-gleichrichter.

LANGUAGE

Asking for information/confirmation

- Could you explain … ?
- I'm not quite sure …
- Does Herr/Frau Böhm want us to … ?
- Does that mean …?
- Do we have to …?
- What's the component in square G5?
- Is that a/an … ?

Giving information/confirmation

- We have to …
- Herr/Frau Böhm wants us to …
- Exactly. / No, he/she doesn't.
- Yes, that's right. / No, it doesn't.
- Yes, we do. / No, we don't.
- It's a/an …
- Yes, it is. / No, it isn't. It's a/an …

9 CHECKING THE COMPONENTS

17 Dennis and Matti are checking that they have got all of the components.

Listen to the dialogue and write down the missing numbers, letters and symbols.

MATTI I think I've got all the components we need now.
DENNIS Good, but we'd better check to make sure. We mustn't forget anything, otherwise it won't work! You tell me what we've got and I'll make sure we've got it.
MATTI OK. We've got a 2 amp fuse and a …[1] amp fuse and a –
DENNIS Just a moment – not too fast! They're pretty small. Right, we've got them. Now what's the next part?
MATTI A transformer with a …[2] output.
DENNIS We've certainly got the transformer. It's so big you can't miss it! How about diodes? How many of them have we got?
MATTI We've got five diodes all the same. They're …[3] diodes.
DENNIS One, two three four, five – OK, we've got them. Now have we got more than one of anything else?
MATTI Yes, we've got two electrolytic capacitors.
DENNIS Both the same?
MATTI No. One's a …[4] electrolytic capacitor.
DENNIS Just a minute. I'm just checking what it says on them. The writing's very small. What's the other one?
MATTI The other one's a …[5] electrolytic capacitor.
DENNIS Yes, we've got both. One's a lot bigger than the other one. Anything else?
MATTI We've got three more things on our list. There's the …[6] adjustable integrated voltage regulator.
DENNIS Here it is and it's got a heat sink to stop it from getting too hot.
MATTI Ah yes. I see. OK, the next component is a …[7] resistor.
DENNIS Another very small part but they're colour-coded to make it a bit easier to identify them. Right, we've got it. And the last part?
MATTI The last part is a …[8] potentiometer.
DENNIS OK, we've got that, too. Now we can get started.

LEARNING

Wie kann ich mein Hörverständnis verbessern?

Je häufiger Sie Englisch hören, desto leichter wird es Ihnen fallen, Muttersprachler zu verstehen. Es gibt viele Möglichkeiten zum Üben:

- englischsprachige Radio- und Fernsehsender (z.B. BBC World, CNN)
- DVDs bzw. Videos im Originalton
- Audiofiles aus dem Internet (z.B. www.bbc.co.uk).

Tipps zum leichteren Verständnis beim Hören

- Machen Sie sich vor dem Zuhören klar, worum es beim Hörtext geht. (Welche Thematik? Worauf lässt der Titel schließen? Gibt es Fotos oder Illustrationen?)
- Versuchen Sie beim ersten Zuhören, den Text sinngemäß zu erfassen, und achten Sie nicht auf Details. (Wenn Sie sich einen Text zum zweiten Mal anhören, können Sie auf Einzelheiten hören.)

10 DESCRIBING ELECTRICAL COMPONENTS

Choose the correct option to explain each component.

1 A **diode** is a device for …
   a converting AC into DC.
   b storing an electric charge.
   c blocking or letting through current.

2 A **resistor** is a device for …
   a limiting how much current flows.
   b converting AC into DC.
   c storing an electric charge.

3 A **capacitor** is a device for …
   a increasing or decreasing voltage.
   b storing an electric charge.
   c protecting a circuit.

4 A **transformer** is a device for …
   a protecting a circuit.
   b increasing or decreasing voltage.
   c storing an electric charge.

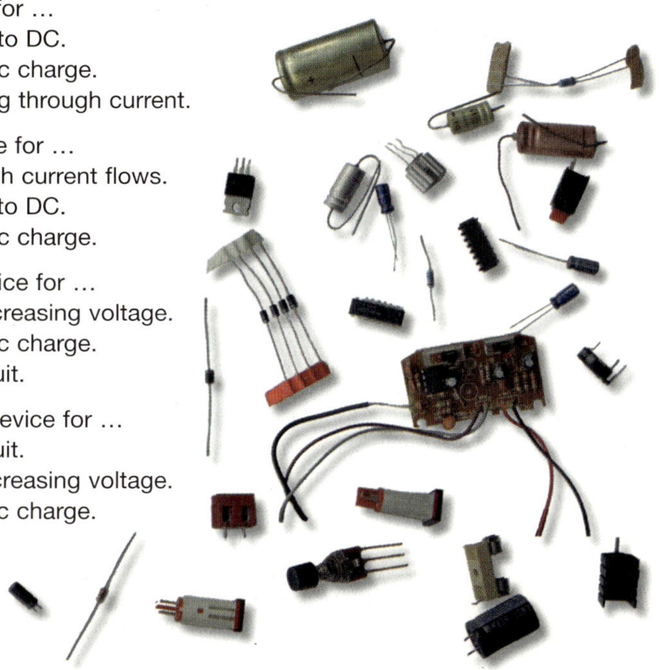

11 DEFINITIONS

Work with a partner:

a Use the table to make definitions.

Example: *A*

A mains tester				conducting heat away.
An oscilloscope				converting AC into DC.
A DMM				joining metals and wires using heat.
A test light		a device		showing voltage against time.
A soldering iron	is	an instrument	for	removing solder.
A desoldering pump		a tool		checking for a voltage of 230 V.
A rectifier				checking for a low voltage.
A heat sink				measuring amps, volts and ohms.

b Now define the following components in your own words:

1 fuse 2 battery 3 photovoltaic cell

12 UNDERSTANDING TECHNICAL DATA

Matti writes up the experiment for his training officer in Finland.

Read the report, then match the five oscilloscope readings (a–e) to the figures 1–5 which Matti refers to in his report.

Trainee: Matti Junkkari
Course: Electrical engineering
Assignment: Making a battery charger
Materials: Circuit diagram, components, printed circuit board
Tools: Soldering iron, pliers, digital multimeter, oscilloscope
Attachment: Circuit diagram of battery charger

Report

First a German trainee and I made a list of the components we needed. We were told to assemble the battery charger step by step and test each component separately before adding the next one.

We began by putting the transformer on the circuit board and testing its output voltage with a DMM. The reading was 12.09 V AC **(figure 1)**. Then we tested the output again with an oscilloscope. It showed us that the AC voltage had two peaks of +17 V and –17 V **(figure 2)**. The DMM had only shown us the rms value.

After that we added the full-wave bridge rectifier and checked the output with the oscilloscope. We saw a fully rectified waveform on the screen (**figure 3**). It was a pulsating DC voltage with peaks of +15.6 V. The voltage was lower because of the voltage drop of 1.4 V across the two diodes.

We then smoothed the DC voltage by adding a capacitor to the circuit. With the oscilloscope we found out that the voltage was still +15.6 V DC. It was no longer a pulsating voltage but it had some ripple **(figure 4)**. We eliminated the ripple by adding the voltage stabilizer after the capacitor and checked the voltage again with the oscilloscope.

There was another voltage drop of about 2 V but the waveform showed a constant DC voltage of 13.5 V **(figure 5)**. Finally, we added a diode to block a reverse voltage from the battery.

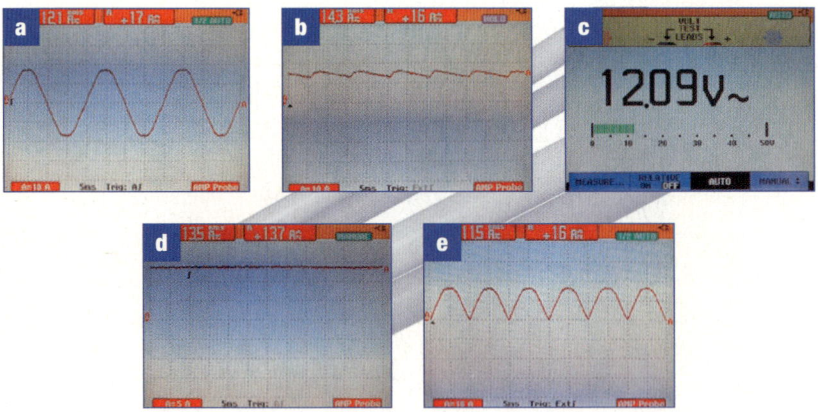

13 UNDERSTANDING THE READINGS

Look at the readings (a–e) on the previous page.

1 Is the multimeter reading in picture c AC or DC? How can you tell?
2 Which component in the circuit changes the waveform from how it is in picture a to how it is in picture e?
3 Which component in the circuit changes the waveform from how it is in picture b to how it is in picture d?

14 MEDIATION

Summarize the report in German for Dennis Lohman's training officer.

15 THE ASSEMBLED BATTERY CHARGER

Dennis and Matti have built the battery charger according to the circuit diagram.

a Look at the picture and identify the components of the battery charger (1–10). The names are all in the box. There are two words too many.

b Now translate the components into German.

10 µF capacitor ■ transistor
2,200 µF smoothing capacitor
220 Ω resistor
4 diodes (full-wave bridge rectifier)
4.7 Ω potentiometer ■ valve ■ fuse
integrated voltage regulator
output diode ■ output terminals ■ transformer

GRAMMAR

Modal verbs: *must* and *can*

Must und *can* sind unvollständige Verben. Bei *must* werden die fehlenden Formen mit *have to* und bei *can* mit *be able to* gebildet.

Infinitiv	**to be able to**	***können***
Present	I can/can't	*ich kann / kann nicht*
Past	I could/couldn't	*ich konnte (nicht) / könnte (nicht)*
Future	I will/won't be able to	*ich werde (nicht) … können*

Infinitiv	**to have to**	***müssen***
Present	I must	*ich muss*
	I mustn't	*ich darf nicht*
	I don't have to	*ich muss nicht*
	Do I have to … ?	*Muss ich … ?*
Past	I had to	*ich musste*
	I didn't have to	*ich musste nicht*
	Did I have to … ?	*Musste ich … ?*
Future	I will/won't have to	*ich werde (nicht) … müssen*

16 PRACTICE

Matti is talking to another apprentice electrician, Josh, from Cardiff, Wales.

Use the Grammar box to complete the dialogue.

JOSH Hi Matti. How are things?

MATTI Great! I'm learning a lot at the company but they start early here.

JOSH What time … (*musst du*)[1] start work in the mornings?

MATTI At 7.30.

JOSH 7.30! That's terrible. We … (*müssen nicht*)[2] start work so early at my company but we … (*dürfen nicht*)[3] be late! My boss goes crazy if we're late. Why didn't you phone yesterday?

MATTI I … (*konnte nicht*)[4] phone because I didn't have a phone card. I've got one now so I … (*kann*)[5] phone any time I want.

JOSH Good. It's my birthday in two weeks' time and you're invited. … (*Kannst du*)[6] come?

MATTI I'm not sure.

JOSH Don't worry. You … (*musst nicht*)[7] buy me a present. You … (*könntest*)[8] just send me a card!

MATTI It's too early to say now but I … (*werde können*)[9] tell you next weekend, all right?

JOSH All right. Sorry, but I … (*kann nicht*)[10] stay on the phone any longer. I really … (*muss*)[11] get back to work!

MATTI OK. Bye.

JOSH Bye.

CULTURE

Grußkarten

Die Briten sind große Freunde von Grußkarten zu allen Gelegenheiten: Geburtstag, Weihnachten, Neujahr, Geburt eines Kindes, Abschiedsfeier usw. Auch Freunde oder Kollegen, die sich jeden Tag sehen, tauschen *birthday cards* und *Christmas cards* aus.

LM117/217
LM 317

1.2 V TO 37 V VOLTAGE REGULATOR

- OUTPUT VOLTAGE RANGE: 1.2 TO 37 V
- OUTPUT CURRENT IN EXCESS OF 1.5 A
- FLOATING OPERATION FOR HIGH VOLTAGES
- COMPLETE SERIES OF PROTECTIONS: CURRENT LIMITING, THERMAL SHUTDOWN AND SOA (SAFE OPERATION AREA) CONTROL

DESCRIPTION
The LM117/LM271/LM317 are monolithic integrated circuit in TO-220, TO-220FP, TO-3 and D²PAK packages intended for use as positive adjustable voltage regulators.
They are designed to supply more than 1.5 A of load current with an output voltage adjustable over a 1.2 to 37 V range.
The nominal output voltage is selected by means of a resistive divide only, making the device exceptionally easy to use and eliminating the stocking of many fixed regulators.

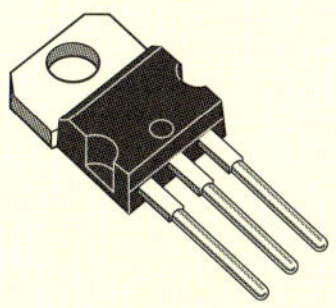

TO-220

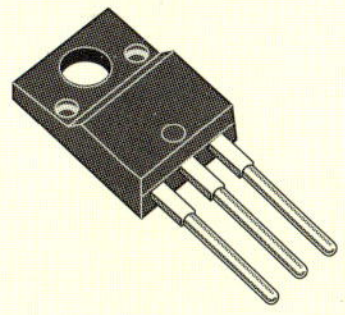

TO-220FP

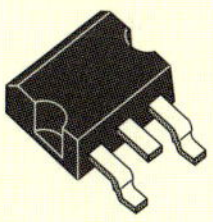

D²PAK

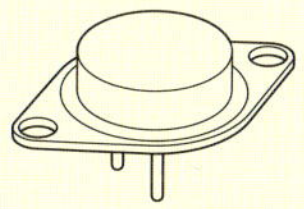

TO-3

1 UNDERSTANDING A DATA SHEET

Study the data sheet, then answer the questions.

1 What are the maximum and minimum voltages produced by this regulator?
2 Is the maximum output current more or less than 1.5 amps?
3 How do you select the desired output voltage?
4 What do you have to mount the voltage regulator on? Why?
5 Why would the voltage regulator heat up?

2 MEDIATION

Summarize the data sheet in German for a colleague who does not speak English.

UNIT 9
Measuring and testing, 2

1 WARM UP

a **What faults can you see in the pictures?**

b **How can you repair them?**

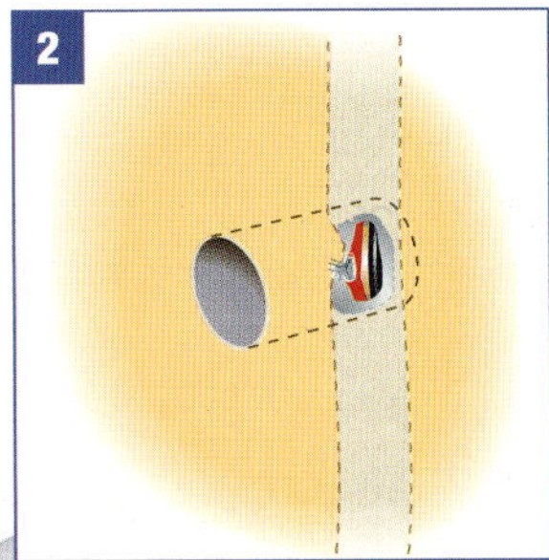

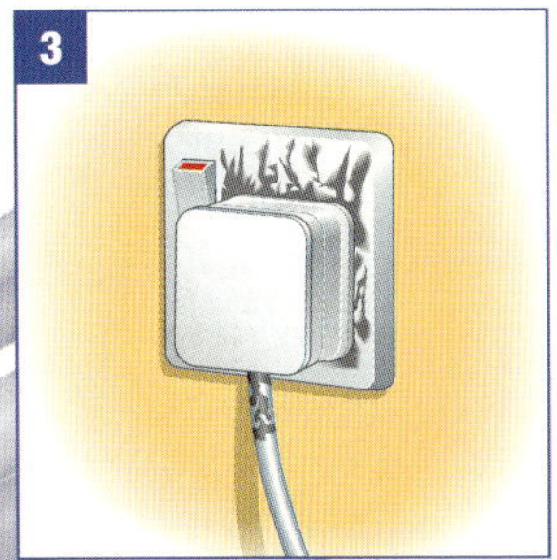

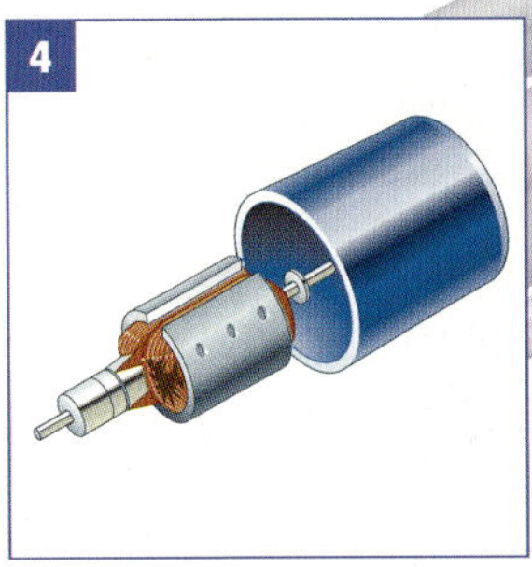

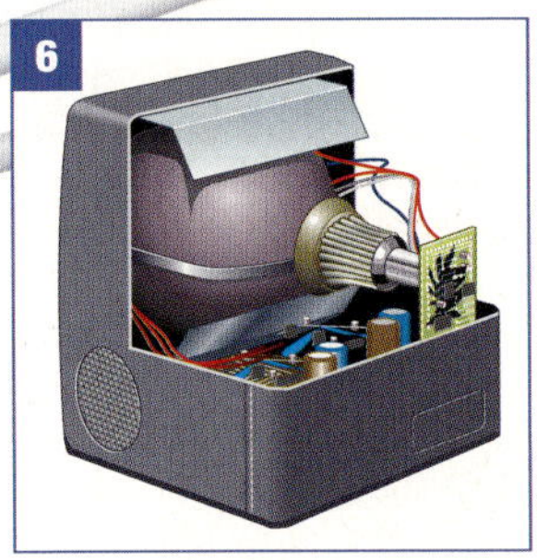

2 A WASHING MACHINE BREAKDOWN

18

Suha Kim is having problems with her washing machine. She phones Electro Tech, a company which repairs domestic appliances.

Read or listen to the dialogue and decide if the statements below are true, false or not mentioned in the text. Correct the false statements.

1 Suha Kim has got a problem with her washing machine.
2 The machine doesn't wash her clothes properly.
3 Jürgen Götz doesn't know yet what has caused the fault.
4 Suha Kim will probably have to buy a new washing machine.
5 Jürgen Götz has to repair two other washing machines tomorrow.
6 Suha Kim gives him all the information he needs about the machine.

JÜRGEN GÖTZ Hallo. Jürgen Götz, ElektroTech. Was kann ich für Sie tun?

SUHA KIM Oh … excuse me, do you speak English? Only, I don't speak German very well.

JÜRGEN GÖTZ Certainly, madam. How can I help you?

SUHA KIM My name is Suha Kim. My washing machine has broken down.

JÜRGEN GÖTZ I see. When you say that it has 'broken down', what do you mean exactly?

SUHA KIM Well, it washes, but it doesn't spin. I've pressed the start button several times and tried some other buttons as well, but nothing happened. What's wrong with it, do you think?

JÜRGEN GÖTZ Well, it could be a fault in the timer control, but there are other possibilities as well. A washing machine that doesn't spin is one of the most common problems we get.

SUHA KIM Can it be fixed easily? The machine is already eight years old: do you think it's worthwhile having it repaired or is it more sensible to replace it?

JÜRGEN GÖTZ Hmm, I can't really answer that one over the phone. It may be quite easy to repair the machine. I can't say for sure until I've looked at it.

SUHA KIM Could you come as soon as possible, please? It's really awkward not having a washing machine, you know, so I need it to be fixed quickly.

JÜRGEN GÖTZ I understand. Let me see … I've got two calls to make tomorrow morning, but I could come to you after lunch, about 2 o'clock.

SUHA KIM That would be fine.

JÜRGEN GÖTZ OK, now I need the model and product code.

SUHA KIM Sorry, I don't know the product code, but I can give you the model. It's a …

CULTURE

Haushaltsgeräte – *domestic appliances*

Haushaltsgeräte sind auf der ganzen Welt ziemlich gleich, würde man meinen. In Wirklichkeit jedoch weisen sie von Land zu Land viele Unterschiede auf.

Während in **deutschen** Einfamilienhäusern Waschkeller eher die Norm sind, sind sie in **Großbritannien** fast unbekannt. Britische Häuser sind in der Regel recht klein und eng und haben selten Keller, deshalb muss die Waschmaschine fast immer in der Küche untergebracht werden. Folglich sind kompakte Waschtrockner bei den Briten beliebt. Kleinere Kühlschränke und Kochherde werden aus Platzgründen ebenfalls bevorzugt.

In den **USA** und in **Australien** herrschen ganz andere Wohnverhältnisse. Da die Häuser und Wohnungen sehr großzügig gebaut sind, gibt es jede Menge Platz für Haushaltsgeräte – und die Geräte sind auch größer und zahlreicher. Doppeltürige Kühlschränke, so groß wie kleinere Kleiderschränke, sind in diesen Ländern keine Seltenheit. In Australien ist es außerdem durchaus üblich, einen *beer fridge* – einen Zweitkühlschrank eigens für Bier – zu haben.

Für Wäschetrockner hingegen besteht in Australien relativ wenig Nachfrage, da das Wetter zum Wäschetrocknen geradezu ideal ist – die gute alte Wäscheleine im Garten ist störungsfrei und verbraucht keinen Strom!

3 WORKING WITH WORDS

Find the expressions 1–8 in the dialogue on page 87 and the sentences on page 96. Choose the best definition for each word.

1 spin
  a behave stupidly
  b turn
  c react
2 several
  a many
  b few
  c some
3 fault
  a malfunction
  b problem
  c development
4 as well
  a too
  b neither
  c luckily
5 common
  a difficult
  b frequent
  c expensive
6 sensible
  a advisable
  b economical
  c practical
7 awkward
  a inconvenient
  b expensive
  c unusual
8 probably (page 86)
  a definitely
  b almost certainly
  c apparently

4 TROUBLESHOOTING WASHING MACHINES

Now complete this text using the expressions from exercise 3.

There are …[1] reasons for a washing machine not to …[2]. One …[3] problem is water in the tub. That's why it is …[4] to check if the machine has drained the water. If it has, the …[5] may be a broken motor or timer. If they are OK, it is …[6] the speed control motor which is faulty. This is an …[7] repair because it requires specialist equipment. Of course, you need technical expertise …[8].

5 READING A TROUBLESHOOTING GUIDE

The text on the opposite page is taken from a troubleshooting guide for a programmable logic controller (PLC) in a washing machine. Read the guide and match the explanations (a–g) to the indicator lights.

X5 — a The solenoid valve is open.
Y0 — b The clutch is engaged.
Y1 — c The brake is disengaged.
Y4 — d The water level is low.
Y5 — e The drain is closed.
Y12 — f The wash motor is in forward gear.
Y27 — g The wash motor is in reverse gear.

TROUBLESHOOTING GUIDE

1 **Machine does not fill**

Check the LED input indicator light corresponding to water level (X5 for low) and the LED output lights corresponding to the drain (Y27) and the water valve (Y12 for HOT).
X5 should be OFF indicating that the required level has not been reached, Y27 should be ON indicating that the drain is closed, Y12 should be ON indicating that the solenoid valve is open.
If X5 is ON, the water level switch may be broken or out of adjustment.
If Y27 is ON and the drain is open, check the drain solenoid valve and the air supply to the drain cylinder.
If Y12 is ON, check the voltage at the water solenoid valve, check the valve to see whether it is stuck closed.

2 **Wash motor will not start**

Check the LED output indicator lights corresponding to the disk brake (Y0), wash motor (Y4 FWD, Y5 REV) and clutch (Y1).
Y0 should be ON indicating that the brake is disengaged. Y4 or Y5 should be ON indicating that the wash motor contactor has been energized. Y1 should be ON indicating that the clutch has engaged.
Check the clutch. Check for sufficient air pressure. Check for leaks from clutch exhaust valve.
Check the motor starter and heater coils.
Check the brake.
Check the door switch circuit.

6 WORKING WITH WORDS

Match the following words from the text with the definitions. Use a dictionary if necessary.

1	input	a	a hole through which liquid is removed
2	indicator	b	what comes out of a machine
3	output	c	setting an instrument etc. correctly
4	drain	d	a device which controls the flow of liquid
5	valve	e	a device which gives information about how a machine is working
6	adjustment	f	when a liquid or gas escapes from a container
7	brake	g	what goes into a machine
8	clutch	h	a device which stops something from moving
9	leak	i	a device which connects a motor to the machine it drives

7 PAIRWORK: TROUBLESHOOTING

Work in pairs.

Student A: Turn to file 7 on page 128.
Student B: Turn to file 16 on page 131.

8 TALKING ABOUT PRODUCT SPECIFICATIONS

Work with a partner. Use the zigzag dialogue to role-play a phone call between a customer and an electrician. The expressions in the Language box will help you. When you have finished, swap roles and do the role play again, using the other washing machine model (opposite).

Customer	Electrician
Sie interessieren sich für (Waschmaschinen-Modell, S. 91), aber Sie möchten mehr Informationen haben.	
	Sie werden Ihr Bestes tun, um die Fragen zu beantworten. Was möchte er/sie wissen?
Was ist die maximale Schleudergeschwindigkeit dieses Modells?	
	Beantworten Sie die Frage mit Hilfe der Produktinformationen (S. 91).
Wie ist der Stromverbrauch/ Wasserverbrauch?	
	Beantworten Sie die Frage.
Wie ist das Stromverbrauch-Leistungs-Verhältnis?	
	Beantworten Sie die Frage. Möchte er/sie noch etwas wissen?
Sie haben noch eine Frage: Wie hoch ist der Geräuschpegel beim Waschen und beim Schleudern?	
	Beantworten Sie die Frage.
Bedanken Sie sich für die Auskunft.	
	Hat er/sie noch eine Frage?
Nein, danke. Sie haben jetzt alle Informationen, die Sie brauchen. Bedanken Sie sich nochmals.	
	Erwidern Sie den Dank und verabschieden Sie sich.

LANGUAGE

Product enquiries

Offering help
How can I help you?
What would you like to know?
Which model do you have / are you interested in?

Asking for information
I'm interested in …
I've got a few questions about …
What can you tell me about … ?

Giving information
That model is rated …
The maximum speed / energy consumption / noise level is …

Concluding
Was there anything else?
Yes, just one more thing.
No, that was it.
Thanks for your help.
You're welcome.

Product information for freestanding home washing machines

Model number	Electrolux 1835	Electrostar 1260
Spin speed maximum rpm	1800	1200
Water litres	43	49
Electricity kWh	1.02	1.06
Voltage V/Hz	230/50	230/50
Fuse protection amps	13	13
Energy performance EN directive 60456	A+	C
Noise level dB re 1pw		
• Washing	48	56
• Spinning	68	73

EN directive 60456: more information for the consumer

The uniform energy label informs shops and consumers about the energy consumption, washing and spinning effectiveness. Through the gradation into efficiency classes from A to G you can see at a glance what kind of appliance you are dealing with.

A or B: A shining example of efficiency, washing and spinning effectiveness
C, D, E: Represents a good to satisfactory 'middle range appliance'
F and G: An appliance at the bottom of the comparison scale

The gradation into classes for energy consumption, washing and spinning effectiveness is based upon values measured in accordance with a standardized procedure in the '60°C cottons' standard wash programme.

9 READING THE SERVICE MANUAL

Use words from the box to complete the instructions.

cabinet ■ clamps ■ installed ■ metal screws ■ motor ■ motor shaft ■ prongs ■ tie ■ transmission ■ unplug

Replacing a washing machine motor coupling

1 Start by removing the ... .
2 Remove the pump from the ... .
3 ... the pump out of the way after getting it off the motor shaft if needed.
4 ... the motor wire harness.
5 Unscrew the two 5/16 sheet ... holding the mounting clamps to the motor.
6 Release the thumb ... and detach the motor from the transmission. Now you can replace the three pieces of the motor coupling.
7 Place either of the two new white couplings on the transmission shaft with the ... facing towards the motor.
8 Place the black rubber union over the prongs on the coupling you just ... .
9 Place the other white coupling on the motor shaft, prongs facing toward the ... .
10 The ... can now be re-installed.

Source: www.applianceaid.com

10 READING TECHNICAL TEXTS

a **Before you read the text on the opposite page, look at the title and headings. Which of the words from the box do you expect to find in the text?**

armature ■ coil ■ commutator ■ download ■ fibre optics
permanent magnet ■ placement machine ■ repel ■ rotor ■ stator

b **Match each motor below (A–C) to the appropriate section of the text (1–3).**

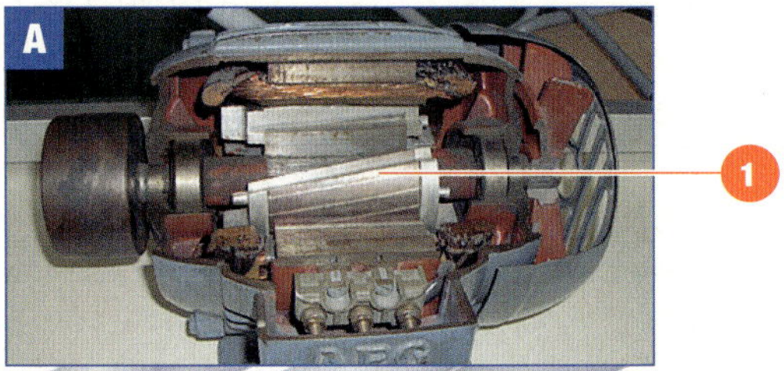

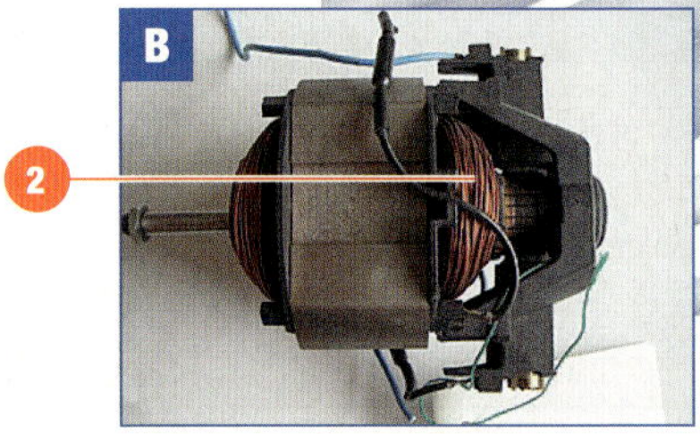

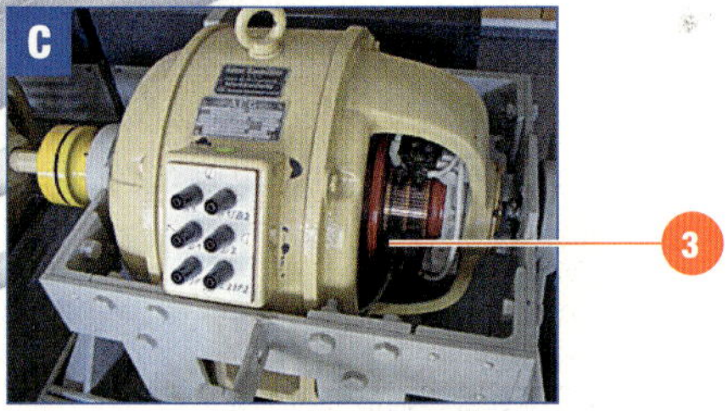

c **Now read the text quickly and say what the numbers in the illustrations refer to.**

LEARNING

Lesestrategien

1 **Vorbereitung auf das Lesen**
Bevor Sie mit dem Lesen eines schwierigen Textes beginnen, vergegenwärtigen Sie sich, was Sie zu diesem Thema bereits wissen.

2 **Erfassen des Sinnzusammenhanges (*Skimming*)**
Verschaffen Sie sich einen Eindruck vom Text, indem Sie diesen zunächst – ohne ein Wörterbuch zu benutzen – rasch überfliegen.

3 **Informationsbezogenes Lesen (*Scanning*)**
Nach dem ersten Lesen eines Textes merkt man häufig, dass nur ein Teil des Textes interessant ist. Konzentrieren Sie sich nur auf die Informationen, die Sie benötigen.

4 **Kontextbezogene Worterschließung**
Versuchen Sie den Sinnzusammenhang (also ganze Sätze anstelle einzelner Wörter) zu verstehen. Meistens erschließt sich die Bedeutung eines Wortes aus dem Kontext des Satzes.

5 **Detailliertes Textverständnis (*Close reading*)**
Wollen Sie einen schwierigen technischen Text in all seinen sprachlichen Differenzierungen verstehen, benutzen Sie ein Wörter- und/oder ein Fachwörterbuch. Dadurch erweitern Sie sowohl Ihren passiven als auch Ihren aktiven Wortschatz.

MENU »

Know it All

Electric motors

The basic scientific principles behind the electric motor are that like poles repel and unlike poles attract, and that electric current flowing through a conductor produces a magnetic field that has both a north and a south pole. An electric motor is essentially a rotating electromagnet. As the electromagnet rotates, electrical energy is transformed into mechanical energy, which can then perform work.
There are many different designs for electric motors, but these are the three basic types:

1 ***DC motor***
In a DC motor, the electromagnet (the **armature**), rotates within a permanent magnetic field created by two opposing magnets (the **exciter field**). The interaction of the north and south pole of the armature and the north and south poles of the exciter field produces a torque, which rotates the armature. A device called a **split ring commutator** reverses the magnetic poles of the armature as it rotates. Carbon **brushes** 'brush' against each half of the split-ring commutator and conduct the electrical energy to the armature. The reversal of the armature poles as they interact with the poles of the exciter field alternately pushes and pulls the armature around.

2 ***AC motor***
A typical AC motor consists of an outside stationary **stator** having coils supplied with AC current to produce a rotating magnetic field, and an inside rotor attached to the output shaft. The **rotor** and shaft are given a torque by the rotating field.
In an **induction AC motor** the rotor rotates more slowly than the magnetic field produced by the electrical supply. Often the rotor is made up of two rings, one at either end, connected by solid bars. Washing machines, dishwashers, etc. use induction motors.
In a **synchronous AC motor** the rotor rotates at the same speed as the rotating magnetic field. This type of motor can also be used as an alternator. Very large synchronous motors are used in the French high-speed train, the TGV.

3 ***Universal motor***
Universal motors may use AC or DC supply current. Unlike DC motors, universal motors have to use an electromagnet to create the stator field, because the field has to reverse its polarity every half cycle of the AC current.
The advantage of the universal motor is that it has the useful characteristics of a DC motor, e.g. high starting torque, high speed and very compact design, but can be used with an AC power supply. This makes them useful for appliances such as blenders, vacuum cleaners, and hair dryers where high-speed operation is desired.

Sources: www.bergen.org and http://en.wikipedia.org

11 WORKING WITH WORDS

Find the English words in the text above.

1 (sich) abstoßen
2 (sich) anziehen
3 Drehmoment
4 (sich) drehen
5 Bürsten
6 Spulen
7 Polarität
8 Hochgeschwindigkeits...

12 MEDIATION

Read the text on page 93 again, using a dictionary if necessary. Explain in German what an electric motor is, and what the main kinds of electric motor are, according to the text.

13 TESTING MOTORS

a **In groups, discuss the steps you would take to test a motor.**

b **Put the following steps for testing a motor into the correct order.**

1. To test continuity, set the multimeter on the RX1 scale to measure resistance.
2. If it tests faulty, replace the motor with one of the same type, power rating, and size.
3. Attach one of the multimeter's probes to the motor's common lead, usually white.
4. Make sure that the power wires are disconnected from the motor.
5. Attach the other probe in turn to each of the other wires on the motor. The probe will be checking to see if it can measure the multimeter's input signal at the output. A low or moderate reading (in ohms) means the component is OK.
6. Remove the motor from the appliance if it is not too difficult.

GRAMMAR

Adjectives and adverbs

1. Repairing the washing machine was an **easy** job.
2. I need it to be fixed **quickly**.
3. It's **really** awkward not having a washing machine.
4. Can it be fixed **easily**?
5. I don't speak German very **well**.

- Man benutzt Adjektive, um Personen oder Sachen näher zu beschreiben. (1)
- Adverbien beziehen sich auf Verben (2, 4, 5) oder Adjektive. (3)
- Adverbien haben in der Regel eine andere Form als Adjektive. Man fügt normalerweise die Endung *-ly* an das Adjektiv an (2, 3). Endet das Adjektiv auf *-y*, wird die Endung zu *-ily*. (4)
- Das Adverb von *good* lautet *well*. (5)

14 PRACTICE

Complete the advice for service technicians, using the adjectival or adverbial forms of the words in brackets.

DEALING WITH CUSTOMERS

- Communicating ... (effective)[1] with customers is an ... (important)[2] part of the service technician's job.
- Always answer the customer's questions ... (polite)[3] and ... (patient)[4], even if you think the answers are ... (obvious)[5].
- You should try not to use too much ... (technical)[6] language in your answers as many customers will not ... (easy)[7] understand complicated explanations.
- If you think that it will be too ... (expensive)[8] to fix the appliance, and the customer should ... (real)[9] buy a new machine, you should say so: customers ... (usual)[10] appreciate ... (honest)[11] answers, even if the answer is not the one they wanted!

Why won't we read the manual?

Caroline E. Mayer

It's an old joke, but a spokesman for Gateway swears it's true – that the computer manufacturer's technical helpline once got a call from a new purchaser complaining that her teacup kept slipping out of the computer's cup holder. The reason: the 'cup holder' was the CD-ROM tray!

Americans buy the most sophisticated computers, the coolest digital cameras, the most advanced automobiles, the most versatile cell phones and handheld organizers, and then they forget or refuse to read the manual.

Marketing experts, customer service consultants and corporate executives all complain that consumers don't read instruction manuals. In fact, manufacturers and retailers say that their customers often return perfectly good items because they just can't make them work.

'Why doesn't my washing machine start?' (Is the lid closed?)

'Why won't my printer work?' (Is it plugged in? Is the ink cartridge properly installed?)

These are among the 'frequently asked questions' (and answers) we find in manuals. But they are also the kinds of questions which manufacturers' free helplines have to deal with. Americans won't read cell phone manuals, kitchen appliance manuals or DVD manuals. They have them all in the cupboard – but they haven't read a single one. They want something that they can take out of the box and just plug in and then it works.

Over the last ten years, there has been a dramatic increase in the number of features, functions and benefits of many products, but as that complexity has increased, consumers have become less willing to sit down and read through complicated manuals.

For that reason, leading electrical appliance producers have started to design products which don't need user manuals. In Europe, for example, a leading appliance company is selling a dishwasher that comes with special software to communicate with the user. In a digital display, the machine first asks which language the user would prefer – French, Dutch, English, Italian or German. Then it asks how many dishes are to be washed and what kind. Pots and pans? China? Crystal? The answers will determine the water temperature and volume and whether the machine needs to filter out hard minerals.

What lesson can we learn from this? You can learn to control technology – or it will control you.

Abridged from: Washington Post, May 26, 2002

1 UNDERSTANDING THE TEXT

Summarize the most important aspects of the text in three to four sentences.

2 DISCUSSION

Discuss these questions with a partner or in a group.

1 Do Germans have the same problem with user manuals?
2 Discuss what could be done to solve the problem.
3 Write guidelines for manufacturers telling them how to write a perfect manual.

UNIT 10
International contacts

1 WARM UP

Look at the cartoon and answer the questions.

1 What do you see in the cartoon?
2 With a partner, find a caption for the cartoon. Which pair has the best caption?
3 What is the cartoonist trying to say?

2 UNDERSTANDING TRADE LITERATURE

Read the text on standards on the opposite page and answer the following questions in German.

1 Why are standards important, according to the text?
2 What standardization organizations are mentioned? Give their full names and explain their areas of competence.
3 How has ISO developed standards? Why?
4 Which important achievement of the IEC is mentioned?
5 What particular problem does the IEC have in developing standards?
6 Is it becoming easier to make standards? Why (not)?

Standards and the consumer – partners for a better world

Safety standards protect life and property. Their application creates safer homes and workplaces and improves the quality of life for everyone. Trading standards are doorways which enable goods to enter other countries and trade to take place on equal terms.

A world without standards would be inefficient and chaotic. That is what rail travel was like in early nineteenth century Britain: each railway company chose its own gauge for its tracks. This meant that locomotives had to be built to different specifications and travellers had to change trains in mid-journey. Finally, in the middle of the nineteenth century, the gauge was set at 4 feet 8.5 inches for the whole country.

This early example of standardization predates the setting up of the International Organization for Standardization (ISO) by more than 100 years. This is the body which consumers have to thank for many of the standards they take for granted.

ISO has been developing standards since it was founded in 1947 and it has always operated by consensus: providers of goods and services have more to gain by following an agreed set of standards than they have to lose. The voluntary nature of standards is also what has made them valuable and workable for consumers, who are more willing to accept regulations which are made for reasons of safety or compatibility than rigid external directives.

For standards to be accepted, the working committees of ISO have had to take into account differences in climate and culture across developing and industrialized countries, and different structures and overseeing authorities.

Since ISO published its first standard in 1951, it has had to return many times to the question of how to create a consensus on what should be measured, in what way, and to what standard.

The International Electrotechnical Commission (IEC), which was created in 1906, has harmonized worldwide the rules for wiring buildings. It has a particular problem because of the fast-moving nature of the industry in which it operates. No matter how fast it reacts to technology development, the computer industry is always going to be moving faster.

The convergence of telephone, computer, DVD and HD television in multimedia systems has led the IEC, ISO and ITU (International Telecommunications Union) to warn that 'an electronic jungle will result unless this exciting rush of development is channelled productively by international standards.'

Today Bluetooth, an industrial specification for wireless personal area networks (PANs), and Wi-Fi, a standard for wireless local area networks (WLANs) are vital for the smooth running of office systems.

Despite 100 years or more of standard-making, the challenges look like being no less complex in the new century as an increasingly integrated world environment – heavily dependent on information technology as its engine – develops at an ever-increasing speed.

Based on an article in the *Financial Times*

3 WORKING WITH WORDS

Choose the best meaning (a–c) for each word (1–10) from the article on page 97.

1 gauge (line 13)
- a standard measurement
- b distance between rails
- c instrument for measuring

2 predate (line 22)
- a happen at many different dates
- b happen at an earlier date
- c exist at a later date

3 take for granted (line 27)
- a to accept
- b get money from
- c take as a fact

4 consensus (line 30)
- a institutions
- b computers
- c agreement

5 voluntary (line 34)
- a done willingly
- b out of the ordinary
- c automatic

6 rigid (line 39)
- a impossible
- b weak
- c inflexible

7 directives (line 40)
- a people who run a company
- b standards
- c instructions

8 take into account (line 43)
- a take money from the bank
- b access an internet provider
- c consider

9 convergence (line 64)
- a becoming different
- b coming together
- c becoming more complex

10 century (line 82)
- a 10 years
- b 100 years
- c 1000 years

4 DISCUSSION: INTERNATIONAL STANDARDS

a **Pick THREE topics from the website below. Why are international standards important in these fields? Write brief notes in English about your opinions.**

b **Work in groups. Discuss the importance of international standards.**

GRAMMAR

Relative pronouns

1 Trading standards are doorways **which/that** enable goods to enter other countries.
2 The ISO is the body (**which/that**) consumers have to thank.
3 Manufacturers are the ones **who** have to follow these regulations.
4 The IEC, **which** was created in 1906, has harmonized the rules worldwide.

- Relativpronomen leiten Relativsätze ein.
- Ein Relativsatz kann notwendig oder nicht notwendig sein. Notwendige Relativsätze sind für das Verständnis des Satzes unbedingt erforderlich (1–3) und stehen ohne Komma, nicht notwendige geben lediglich Zusatzinformationen und werden durch Kommas getrennt. (4)
- In notwendigen Relativsätzen kann das Relativpronomen wegfallen, wenn es Objekt ist, d.h. wenn im Relativsatz ein Subjekt steht. (2)

5 PRACTICE

With a dictionary, translate the following sentences. Use relative pronouns where necessary.

1 Der Artikel, den ich über internationale Standards gelesen habe, ist sehr interessant.
2 Die Geräte unserer Firma, die weltweit verkauft werden, sind von hervorragender Qualität.
3 Geräte, die innerhalb von zwei Jahren kaputtgehen, werden unter Garantie repariert.
4 Unsere Firma sucht jemanden, der auch im Ausland tätig sein möchte.
5 Der Computer, der gestern abgestürzt ist, funktioniert wieder.
6 Interelectronics verkauft Produkte, die weltweit beliebt sind.

6 DEFINITIONS

Pick a word from the box. (Don't tell your partner the word!) Use the grid to make a definition for the word, then read out your definition. Can your partner guess the word?

electrician ■ engineer ■ multimeter ■ plug ■ programmer ■ solar cell ■ standard ■ trainee

It's	someone something	who which	connects converts designs learns measures repairs serves writes	a piece of electrical equipment to the electricity supply. as an official unit of measurement. electrical equipment. instructions for computers. light and heat energy into electricity. machines. a trade from soneone who is already qualified.

7 ELECTRICITY AROUND THE WORLD

a **Describe the picture on the opposite page and explain its message.**

b **Read the text on the opposite page and look at the eight types of plugs mentioned in the text. Then copy the table below and complete it with information from the text.**

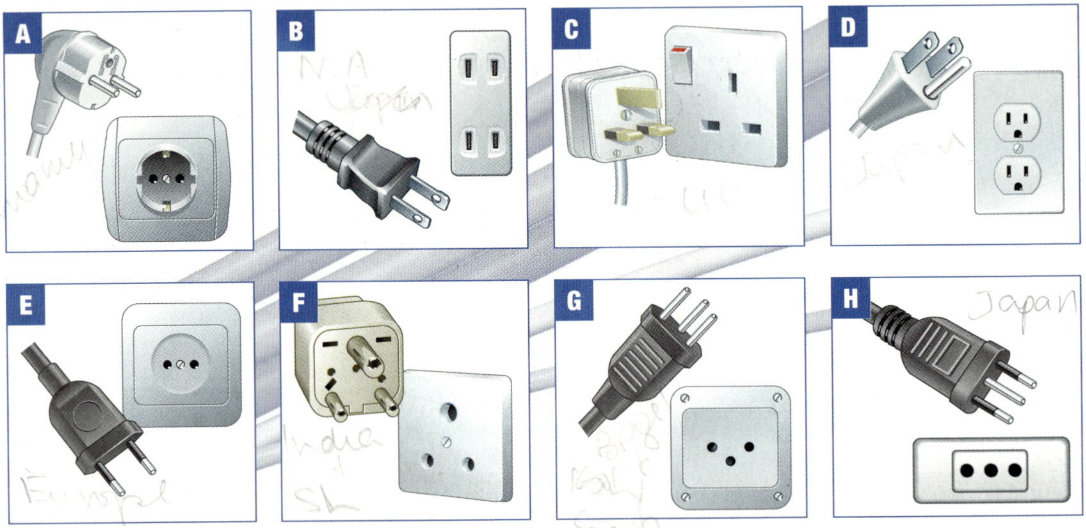

TYPE	SHORT DESCRIPTION	COUNTRY
A	Two round prongs and grounding contacts	Germany

8 WORKING WITH WORDS

Match each word (1–8) from the text on the opposite page to its definition (a–j). There are two definitions too many.

1	mains	a	letter to apply for a job
2	differ	b	difference
3	appliance	c	not be the same
4	disparity	d	take over
5	adopt	e	short piece of metal
6	pin	f	not in one line
7	exclusively	g	very expensive
8	offset	h	limited to a certain group
		i	device for a specific task
		j	AC electricity supply in the house

File Edit View Go Bookmarks Tools Help

http://users.pandora.be/worldstandards/electricity.htm Go

Home Electric FAQ

Why did my hair dryer blow up?

There is no standard mains voltage throughout the world; also the frequency varies from country to country. Plug shapes, plug holes, plug sizes and sockets also differ. These seemingly unimportant differences have some inconvenient consequences: most appliances from one country simply cannot be connected to the wall outlets in another country.

Voltage and frequency

While it is easy to get a plug adapter or a new 'local' plug, in many cases this only solves half the problem, because it doesn't help with the possible voltage disparity. A 110-volt electrical appliance designed for use in North America or Japan will provide a nice fireworks display if plugged into a European socket.

There are two main types of electric systems used around the world, with varying plugs and sockets. Europe and most other countries in the world use 220 to 240 volts at 50 hertz frequency, whereas Japan, the USA and most South American countries use 100 to 127 volts at 60 hertz.

Plugs

Worldwide there are no less than 13 different styles of plugs and wall outlets, because many countries preferred to develop a plug of their own, instead of adopting the US standard. The main types of plug around the world are listed below:

In Germany the so-called 'Schukoplug' is used. It is round, has two round pins and two grounding contacts on the side of the plug. An ungrounded plug with two flat parallel prongs is standard in North and Central America. Almost exclusively in Italy a plug type is used which consists of three round pins which are in line. In the UK and China a plug is used which has three rectangular pins that form a triangle. A plug with two flat parallel prongs and a round grounding pin is used in Japan. Almost exclusively in India and Sri Lanka a type of plug is used that has three round pins arranged in a triangle. In all countries of Europe except the United Kingdom and Ireland a plug is used which is ungrounded and has only two round pins. It is popularly known as the Europlug. In Brazil and Switzerland the official standard plug has three round pins which are offset.

9 PREPARING FOR INTERNATIONAL BUSINESS TRIPS

Standards such as those for plugs are only some of the things to be considered when preparing for an international business trip.

Work in a small group:

1 Make a list of things you have to take into account when going on a business trip abroad. Use the checklist below to start you off.
2 Rank the items in your list in order of importance.
3 Compare ideas with other groups.
4 Make a handout for your company's personnel.

CHECKLIST FOR BUSINESS TRIPS ABROAD

Before going abroad on business, make sure you take the following items into consideration.

- Customs regulations
- Weather conditions in the country you are visiting

LEARNING

Tipps für das Gruppengespräch auf Englisch

Sich zu Wort melden
Can I just say that … ?
Well, if you ask me …

Die eigene Meinung äußern
I think that …
In my opinion, …

Jemand anderen ins Gespräch einbringen
What do you think, Carsten?

Höflich widersprechen
You may be right, but …
I don't think that's quite right.

Nachfragen, ob man verstanden wurde
Do you see what I mean?

Nachfragen, ob man etwas richtig verstanden hat
Do you mean that …?

Auf die anderen Gesprächsteilnehmer reagieren

I see. / Oh, right.	Ach so.
Really?	Echt?
Right.	Ah ja.

10 PREPARING TO LISTEN

Work with a partner. You are going to hear a presentation about cultural differences between Germany, China and the USA. What differences will be mentioned, do you think?

11 A PRESENTATION ON CULTURAL DIFFERENCES

19

a Copy the table below. Then listen to the presentation and complete columns A and B with key words or short sentences.

b Now fill in column C with your own ideas. Use the internet if necessary.

	A China	B USA	C Germany
1 Attitude towards time			
2 Business culture			
3 Company culture			
4 Body language			
5 Technical standards			

12 LANGUAGE FOR PRESENTATIONS

19

Listen again. Which of the expressions 1–11 do you hear?

1 Good morning/afternoon/evening. My name is …
2 I'm going to talk about …
3 Let's move on to …
4 If you have any questions, please feel free to interrupt me.
5 If you have any questions, I'd prefer to take them at the end.
6 Now it's time to talk about …
7 To begin, I'd like to look at …
8 Let me give you an example.
9 Have a look at this photo / graph / bar chart …
10 Before I finish, I'd like to summarize the key points.
11 Well, if there are any questions, I'd be happy to answer them.

13 IT'S YOUR TURN

Prepare a presentation about the key product or service of your company.

1 Make notes about all the important information.
2 Write your presentation. Use expressions from exercise 12.
3 Give your presentation.

14 AT A TRADE FAIR

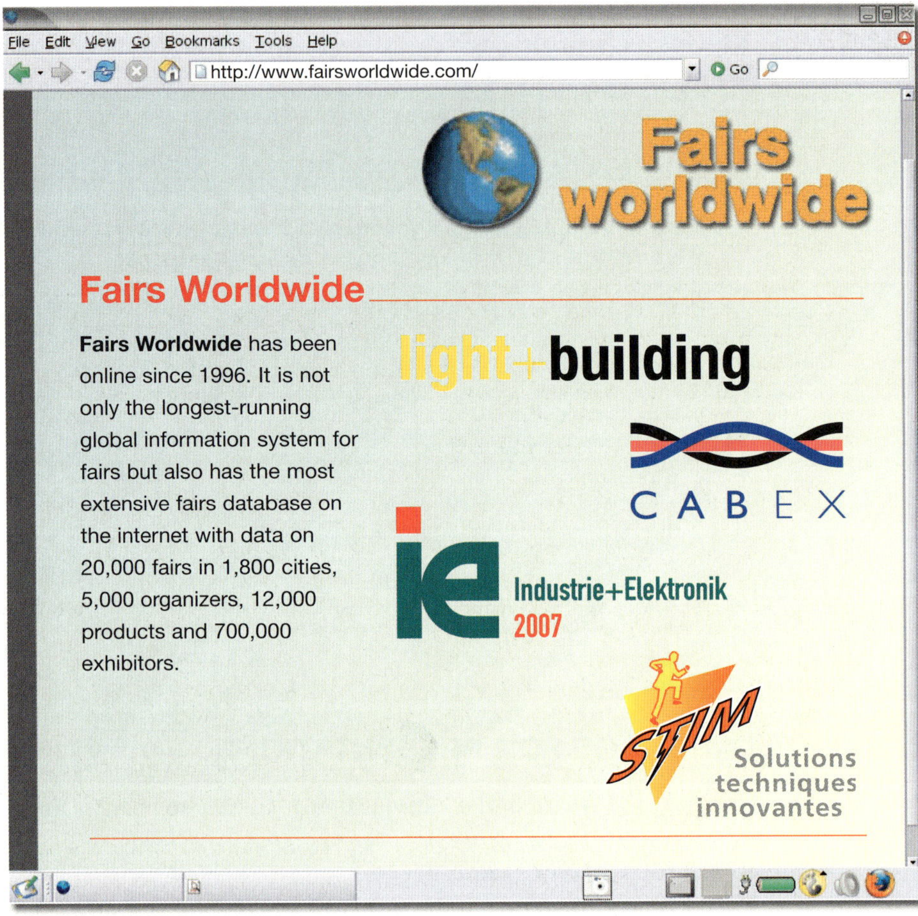

a **What trade fairs are represented by the logos on the website? Name some other trade fairs that are important for your company.**

b **A customer comes to a trade fair and wants to inquire about a product. Re-write the following dialogue to make it polite.**

COMPANY REPRESENTATIVE	Hello. I want to introduce myself. I'm Wolfgang Pulzer from Interelectronics. This is our latest product. We want to launch it at this trade fair.
CUSTOMER	It looks good. Have you got a brochure with technical details?
COMPANY REPRESENTATIVE	No, we haven't. We have only got a leaflet with general information. I can send you detailed info if you want.
CUSTOMER	OK. Take my business card.
COMPANY REPRESENTATIVE	Good. You'll get the brochure next week.
CUSTOMER	Alright.
COMPANY REPRESENTATIVE	Here's my card. If you need more information, I'll help you.
CUSTOMER	Thanks.
COMPANY REPRESENTATIVE	Bye.
CUSTOMER	Bye.

Cross-cultural pitfalls

When you are working abroad, it is important to know about the culture and customs of the country you are in. Failures in communication due to cross-cultural ignorance can cost clients and money.

How good is your cross-cultural knowledge? Do the culture quiz to find out.

1. In Thailand you may not touch a person's …
   a head **b** hand **c** shoulder

2. In Bulgaria nodding means …
   a 'No.' **b** 'Yes.' **c** 'I don't know.'

3. The sign thumb and index finger forming a circle and the other fingers raised is in Japan …
   a a vulgar expression
   b a sign for 'OK'
   c a sign for 'money'

4. The number 4 is avoided because it means death in …
   a Britain **b** Brazil **c** China

5. When you immediately put a business card into your pocket without looking at it, you offend …
   a a French person
   b a Japanese person
   c an American

6. If an important business meeting is cancelled because it is raining, you are in …
   a Paraguay **b** England **c** Finland

7. In Muslim countries you cannot do business …
   a on Fridays
   b on Saturdays
   c on Sundays

8. You should not ask about your client's family before getting down to business in …
   a Italy **b** China **c** Arab countries

1 TEST YOUR KNOWLEDGE

Do the quiz. Use the internet to check any answers you aren't sure about.

2 NATIONAL STEREOTYPES

Look at the cartoon and discuss the following questions.

1 Which nationalities has the cartoonist drawn?
2 What is he trying to say about the characters?
3 Do you agree with the stereotypes? Why (not)?

UNIT 11
On installation abroad

1 WARM UP

Work with a partner.

a Match each site (1–4) with the correct photo (A–D) and electrical installation (a–d).

1 Textile mill, Zambia
2 Car factory, Brazil
3 Circuit board factory, South Korea
4 Brewery, China

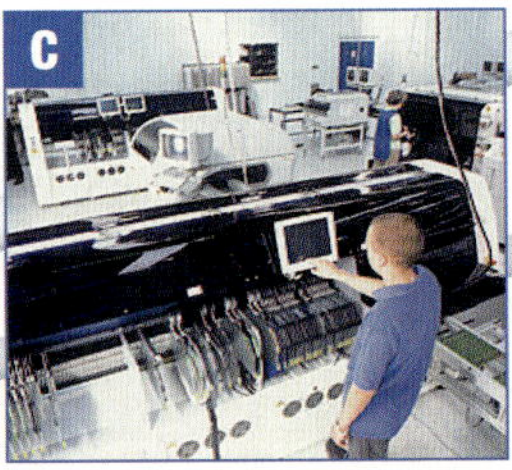

a Install a thermostat in the refrigeration plant.
b Install a component feeder in a placement machine.
c Install a coolant flow monitor for the robotic welding station.
d Install a control panel for a weaving machine.

b Which installation would you find most interesting? Why?

c What other examples of installation abroad can you think of?

2 A PROJECT MEETING

BeerTech, an international supplier of brewing technology, is supplying technical equipment and machinery for a new brewery at Guangzhou in China.

a Read the agenda, then answer the questions below.

1 When will the meeting take place?
2 How many people will be present at the meeting?
3 Who is in charge of the project?
4 Which phase of the project should have been completed by the time of the meeting?
5 Which phase of the project should start soon?

b Now complete the minutes of the meeting with information from the agenda.

AGENDA – GUANGZHOU BREWERY PROJECT MEETING

Venue	Room 204
Date	Wednesday 25 September 20..
Time	14.00–17.00
Chair	William Fallon, project manager
Attendees	Yang Li, Jürgen Koller, Peter Morgan, Lothar Zeidler
Minutes	Irene Urban
Objectives	Project status review

Items

1 Apologies
2 Minutes of last meeting
3 Review of project status:

	Responsibility	Schedule
Preparation	YL, PM	5 May–28 June

- Site preparation
- Reception of technical equipment
- Clarification of installation procedures for local subcontractors

Installation	YL, PM, LZ	1 July–12 Oct

- Installation of brewing machinery and technical equipment
- Installation of monitoring and control systems
- Supervision and quality control of electrical subcontractors

Commissioning	JK, LZ	15–26 Oct

- I/O test
- wet test
- test run
- acceptance test

4 Reporting procedures
Review of reporting procedures to project manager on site and to construction department at BeerTech head office

5 Any other business

MINUTES

Guangzhou Brewery Project meeting 25.09.20..

The project team reviewed the ...[1] of the Guangzhou Brewery Project.

Ms Yang and Mr Morgan reported on progress to date:

Ms Yang reported that the ...[2] of brewing ...[3] and technical equipment was completed by 20 August. Mr Morgan confirmed that the installation of ...[4] systems is running according to schedule.

Mr Koller was introduced to the team. He will be responsible for the ...[5] of the equipment alongside Mr ...[6] They will start work in China on 15 October with the I/O and wet test and complete the final test run and ...[7] test by 26 October at the latest.

Mr Fallon reminded the meeting that all technicians on installation in China must report to the ...[8] on site on a weekly basis. He will report to the ...[9] at head office on a monthly basis and ad hoc (on important events).

3 WORKING WITH WORDS

Match the definitions (a–h) to the expressions (1–8). All of the expressions are used on page 107.

1 venue
2 minutes
3 chair
4 item
5 objective
6 attendees
7 other business
8 agenda

a Additional subjects that people want to discuss at the meeting
b Where the meeting takes place
c The written plan for the meeting
d A record of what was said at the meeting
e A subject to be discussed at the meeting
f The people who will be present at the meeting
g What the meeting should achieve
h The person in charge of the meeting

4 DISCUSSION

Work in a small group. You are going on installation abroad. What practical arrangements will your company need to make for your trip? Make a list.

5 MAKING TRAVEL ARRANGEMENTS

20

Jürgen Koller is preparing to leave for China. He checks his travel arrangements with Will Fallon's assistant, Mary Khor.

Read or listen to the dialogue. Are any of your ideas from exercise 4 mentioned? Which ones?

JÜRGEN Hi, Mary. It's Jürgen. Can you just run through the travel arrangements for Guangzhou with me?

MARY Sure … Your flight departs from London Heathrow at 17.40 on Sunday and arrives at Guangzhou at 17.15 on Monday, that's local time.

JÜRGEN Is that a direct flight, or is there a stopover somewhere?

MARY There's a three-hour stopover in Beijing, where you change planes for Guangzhou.

JÜRGEN OK. Which airline am I flying with?

MARY You're flying with Air China.

JÜRGEN Fine. Is anyone meeting me at the airport when I arrive at Guangzhou?

MARY Yes, Will's going to arrange for someone to meet you at the airport.

JÜRGEN Great, because I don't speak a word of Chinese and I've never been to China before.

MARY Don't worry, Jürgen. It will be fine. Was there anything else?

JÜRGEN I don't think so … Oh, yes – one more thing: will I need my international driving licence?

MARY No, you won't. Our Chinese customer is going to provide a car and a driver for you.

JÜRGEN Good. That will make things a lot easier.

MARY I'll send you an email confirming the arrangements.

JÜRGEN Thanks. Bye, Mary.

6 UNDERSTANDING THE TEXT

Read the dialogue on page 108 again, then answer the questions.

1 Does Jürgen have a direct flight to Guangzhou?
2 How many times has he been on installation in Guangzhou?
3 Who is going to arrange for someone to collect him at the airport?
4 What is he worried about?
5 How is he going to get around in Guangzhou?

GRAMMAR

Talking about the future

1 **I'll send** you an email confirming the arrangements.
2 He **will be** responsible for the adjustment of the equipment.
3 **Is** anyone **meeting** me at the airport?
4 Our Chinese customer **is going to provide** a car for you.
5 Your flight **departs** from London Heathrow at 17.40.

- Das *will-future* (*will/won't* + Grundform) steht für spontane Entscheidungen, Versprechungen und allgemeine Aussagen über die Zukunft. (1, 2)
- Das *Present progressive* wird für feste Verabredungen oder Pläne verwendet. (3)
- Das *going to-future* (*going to* + Grundform) bezeichnet eine zukünftige Absicht. (4)
- Das *Simple present* wird für Fahr- und Zeitpläne verwendet. (5)

7 PRACTICE

Will Fallon has already flown out to China to supervise the project. He phones Jürgen to check his travel plans.

Choose the best future forms to complete the phone call.

WILL Hi, Jürgen. It's Will here. Are you all set for Guangzhou?
JÜRGEN Yes, thanks. Mary has sorted everything out for me.
WILL Good. What time … (your flight / arrive)[1]?
JÜRGEN It … (get in)[2] at quarter past five on Monday afternoon.
WILL Right. I … (let)[3] Mrs Li know.
JÜRGEN Mrs Li?
WILL She's the company rep who … (collect)[4] you and take you to the hotel. Then I … (take)[5] you out for dinner and show you around town, if you like.
JÜRGEN Great!
WILL Oh, and on Wednesday we … (give)[6] the local press a tour of the site. I … (talk)[7] about the brewing process, and I'd like you to explain the commissioning procedure. … (that / be)[8] OK?
JÜRGEN Yes, of course. I … (make)[9] some notes during the flight.
WILL Good idea. Well, have a good trip. I … (see)[10] you on Monday, then.
JÜRGEN Thanks, Will. See you then.
WILL Bye for now.

CULTURE

Englisch unterwegs

Wenn Sie mit dem Flugzeug unterwegs sind, werden Sie feststellen, dass Englisch auf Flughäfen die Sprache Nr. 1 ist. Merken Sie sich folgende wichtige Durchsagen: Worum geht es?

Will passenger Petra Schmidt please return to the check-in desk?
Passengers for flight BA101 may now proceed to gate 20.
Flight BA101 is now boarding at gate 20.
Last call for passengers for flight BA101 at gate 20.
Flight BA101 has been delayed. Please consult the departures board.
Please do not leave your bags unattended in the airport.
Will all persons leave the building immediately via the nearest exit.

8 AIRPORT FACILITIES

Match each airport facility (1–4) with the correct picture and speech bubble(s).

1 Security control
2 Information desk
3 Customs counter
4 Check-in counter

A

B

C
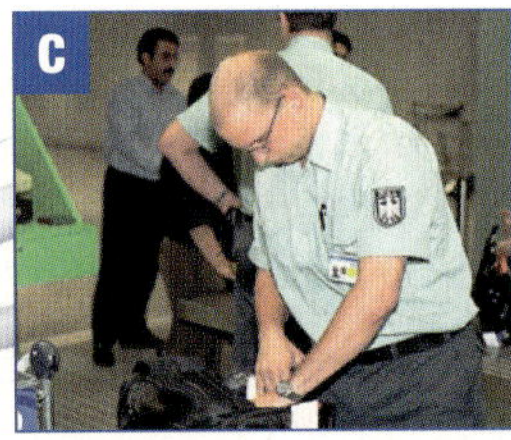

D
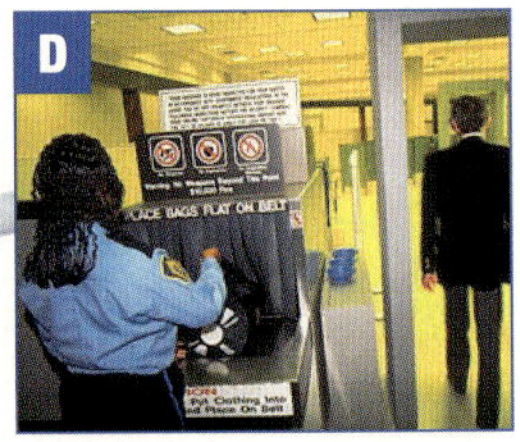

a *Do you have anything in your pockets?*

b *Do you have anything to declare?*

c *How many pieces of luggage are you checking in?*

d *May I see your ticket and passport, please?*

e *What is the purpose of your visit?*

f *Follow the signs for Terminal 4.*

9 EN ROUTE TO GUANGZHOU

21

Jürgen Koller is on his way to China.

Listen to the dialogues. Which of these problems does Jürgen have?

Dialogue 1: information desk
1 He's at the wrong terminal.
2 He can't find the shuttle train.

Dialogue 2: check-in
3 His luggage is over weight.
4 He can't have a window seat.

Dialogue 3: security control
5 The metal detector goes off.
6 He has a sharp metal object in his hand luggage.

Dialogue 4: Guangzhou customs
7 He tries to take too much beer through customs.
8 He hasn't completed his customs declaration card.

10 COMMISSIONING THE PLANT

22

Jürgen Koller explains the commissioning process to a group of Chinese journalists at the brewery.

Read or listen to the dialogue, then answer the questions below.

JÜRGEN Hello, everyone. I'm Jürgen Koller, the technician responsible for the electrical installation. Our task today is to find the perfect settings for the brewing machines and equipment. We've just completed the I/O test successfully.

JOURNALIST I/O test? What's that, please?

JÜRGEN That's the in/out test. We used this laptop with its specially designed software to test if the signals between the PLCs and the motor or pumps of the brewing equipment were OK.

JOURNALIST I see. So you can control the whole plant with your laptop. Is that right?

JÜRGEN That's right, yes.

JOURNALIST Very impressive. So what will happen next?

JÜRGEN The next stage is the wet test.

JOURNALIST The wet test?

JÜRGEN That's a test run with water so that we can see that everything works smoothly ... Well, as you can see on the screen we have to change the settings of the flow meters. At the moment the flow rate is too low and the production output per hour would not meet the requirement of our customer. We also have to adjust the RPM of motor 3 – it's too fast.

JOURNALIST And all this can be done with the laptop and the right software, can it?

JÜRGEN That's right.

JOURNALIST Will the Guangzhou Brewery have to fly you and your laptop out from Britain every time they need to adjust the settings?

JÜRGEN No. We're going to instruct the brewery technicians in the use of this software. Smooth knowledge transfer is part of our company's policy. We think that plant efficiency can only be increased if the employees themselves are able to solve the problems that come up in daily operation after the handover of the plant to the customer.

JOURNALIST That sounds like a very good policy.

JÜRGEN Yes, our customers think so, too ... Well, I can see that the wet test run was successful. All the settings are perfect now. Now we're ready to do a first test run in which real beer is produced. If we're happy with that, we'll progress to the acceptance test later in the week.

JOURNALIST And the acceptance test is ...

JÜRGEN The final commissioning test, which our Chinese colleagues will carry out under our guidance, to make sure that they can operate the system and that they're satisfied with its performance.

1 Which test has Jürgen's team already completed?
2 What adjustments does Jürgen make during the wet test?
3 Who will make adjustments after the handover?

11 WORKING WITH WORDS

Explain these terms from the dialogue on page 111 in your own words.

1 I/O test
2 wet test
3 knowledge transfer
4 test run
5 acceptance test
6 handover

12 PROGRESS REPORT

You are Jürgen Koller. Read the dialogue again, then write an email in German to Lothar Zeidler, the German brewmaster at BeerTech, saying where you are up to with the commissioning process.

13 PLCS

a How much do you know about PLCs? Decide whether these statements are true or false.

1 PLC is the abbreviation for 'programmable logic circuit'.
2 A PLC can store instructions in its internal memory.
3 A PLC is a type of small computer.
4 PLCs can be used to control sensors, switches, temperature indicators, etc.
5 Input/Output arrangements can be included in a PLC.
6 Newer automated systems use relays instead of PLCs.
7 It's cheaper to use relays than PLCs.
8 The automotive industry was one of the first to use PLCs.

b Now read the text below to check your answers. Correct the false statements.

PLC (Programmable logic controller)

A programmable logic controller (PLC) is a digital electronic device which uses a programmable memory for the internal storage of instructions for implementing specific functions, such as sequencing and timing and to control machines through digital or analogue input/output (I/O). In other words, a PLC is a small computer used for automation of real-world processes.

PLCs are connected to sensors, switches, temperature indicators, etc. They can drive any kind of electric motor, pneumatic or hydraulic cylinder, magnetic relay or solenoid. That's why one of the main uses of PLCs is to control machinery, e.g. on factory assembly lines.

The I/O interface may be built into a simple PLC, or the PLC may have external I/O modules attached to a proprietary computer network that plugs into the PLC.

PLCs were invented as less expensive replacements for older automated systems that used hundreds or thousands of relays and cam timers. A single PLC can be programmed to replace thousands of relays. Initially, programmable controllers were used by the automotive industry, where they enabled software revision to replace the rewiring of hardwired control panels.

14 WORKING WITH WORDS

Find the English words in the text on page 112. They are in the same order in the text.

1 programmierbar
2 Speicher
3 Speicherung
4 Befehle
5 Automatisierung
6 Schalter
7 Anzeiger
8 steuern
9 Magnetschalter
10 Montageband
11 Eingabe-/Ausgabeschnittstelle
12 Schalttafel

15 PREPARING A PRESENTATION

Jürgen Koller has to familiarize the Chinese brewery managers with the technical details of the new brewery plant. He uses the text on page 112 to help him to prepare a short presentation on PLCs.

Work with a partner:

a Write the presentation on PLCs for Jürgen. Remember that he is talking to a non-technical audience, so use simple English which can easily be understood.

b Read the advice in the Learning box, then give the presentation to your class.

LEARNING

Wie kann ich meine Sprechfertigkeit verbessern?

Wie man sich im Sprechen ausdrückt, hängt insbesondere vom Sprechanlass ab: Referate, Vorträge und Präsentationen erfordern teilweise andere Fertigkeiten als das Sprechen in Alltags- oder beruflichen Situationen, wenn es gilt, möglichst direkt auf die Aussagen des Gegenübers zu reagieren.

- Formulieren Sie bei einer Rede oder einem Vortrag den ersten und letzten Satz vor, um ein Stocken des Sprachflusses zu vermeiden, z.B.:

Today I'd like to talk about programmable logic controllers …

In conclusion, we can see that PLCs are a cheaper and more flexible alternative to relays.

- Anstatt die ganze Rede vorzulesen, schreiben Sie die wesentlichen Punkte als kurze Notizen auf Karteikarten, die Sie dann während der Rede als Wegweiser benutzen können. Das macht Ihre Intonierung natürlicher und Ihre Rede interessanter.
- Machen Sie Ihre Sätze kurz und aussagekräftig, damit Sie das Interesse der Zuhörer behalten und sich nicht in komplizierten Satzstrukturen verheddern.
- Sprechen Sie langsamer, als Sie in einem normalen Gespräch reden würden, mit kurzen Pausen nach den wichtigsten Aussagen.

16 MEDIATION

You are the junior electrician on an installation project abroad. You are installing new brewery equipment in South Africa. Your boss, the senior electrician, isn't confident in English, so you offer to mediate between him and the manager of the brewery.

Mediate between the senior electrician and the manager.

Electrician	You	Brewery manager
Guten Tag. Ich bin Peter Meier, von der Firma BrauTechnik, und das ist mein/e Assistent/in, [Name]. Wir sind hier, um die elektrische Installation abzuschließen.	?	
		Hello. Welcome to our company. My name is Mr de Vries. Please tell me what is involved in the electrical installation.
	?	
Die elektrische Installation wurde zum größten Teil bereits in unserer Firma gemacht. Unsere Aufgabe ist, die Technik zu überprüfen und optimal einzustellen.	?	
		I think this pre-installation method is very good and effective. What will the testing process involve, exactly?
	?	
Wir werden zunächst den I/O Test und den Nass-Test mit einfachem Wasser machen. Danach wird der eigentliche Testlauf durchgeführt. Anschließend kann die Übergabe erfolgen.	?	
		That sounds fine. Are you going to train our staff so that they can maintain the equipment as well and know what to do in case of a breakdown?
	?	
Ja, natürlich. Es ist unsere Firmenpolitik, technisches Wissen an die Kunden weiterzugeben, damit sie den Herstellungsprozess eigenständig am Laufen halten können.	?	
		Excellent! We're glad to cooperate with your company and are very satisfied with the work you are doing.
	?	
Wir bedanken uns ebenfalls für die gute Zusammenarbeit.		

File Edit View Go Bookmarks Tools Help

http://www.china-window.com/china_business/doing_business_in_china/index.shtml Go

On installation in China: What you should know

Names

Chinese names appear in a different order to Western names. In Chinese names, the family name comes first. For example, in the name Chen Wu, Chen is the family name and Wu is the given name. He should be addressed as 'Mr Chen'. Only family members or close friends use given names.

Greetings

Chinese people usually shake hands very lightly instead of taking the hand firmly and forcefully pumping it, and in China a handshake may last as long as 10 seconds. Another accepted form of greeting is a nod or slight bow. Upon meeting someone, Chinese people lower their eyes slightly as a sign of respect. Staring into the eyes of a Chinese person might make him or her uncomfortable.

'Face'

Face – being respected by one's peers – is very important to the Chinese. Saying or doing anything that causes someone to lose face can instantly destroy a relationship and any business that might result from it. Never insult or openly criticize someone in front of others. Don't make fun of a Chinese person, even if only as a joke.

Meetings

Meetings are very formal in China. A list of your representatives before the meeting including their rank in your company is required. You should also request a similar list. You will be led into a room in which the Chinese are already present. Your team leader should enter first. Teams sit across a table, leaders opposite each other and others seated in descending order of importance.

Small talk will come first. The head of the host team will deliver a short welcome speech, then turn the floor over to the visitors. Your senior team member should speak for your company. When talking, your spokesperson should address the senior Chinese representative.

Conversation

Negative replies are considered impolite in China. Never say 'No.' Instead, you can respond with 'I'll look into that,' 'I'll think about it,' or 'Maybe.' When your Chinese counterparts enthusiastically say 'No big problem,' they usually mean 'There are still problems.'

The question 'Have you eaten?' is like 'How are you?' in Western countries: it's a polite expression that does not require a detailed answer. Simply answer 'Yes,' even if you haven't eaten or just smile and say 'Thank you.'

1 UNDERSTANDING THE TEXT

Read the text, then give five examples of how an electrician on installation in China could accidentally offend his/her customers.

2 CROSS-CULTURAL DIFFERENCES

Think of a country you would like to visit. Use the internet to find out about cross-cultural differences, then write a list of tips for people visiting this country. Be critical: watch out for 'differences' which are really just stereotypes (e.g. 'Germans don't have a good sense of humour').

UNIT 12
Applying for a job

1 WARM UP

As a qualified electrician or electronic technician with a working knowledge of English, you can apply for a job in any country where English is spoken.

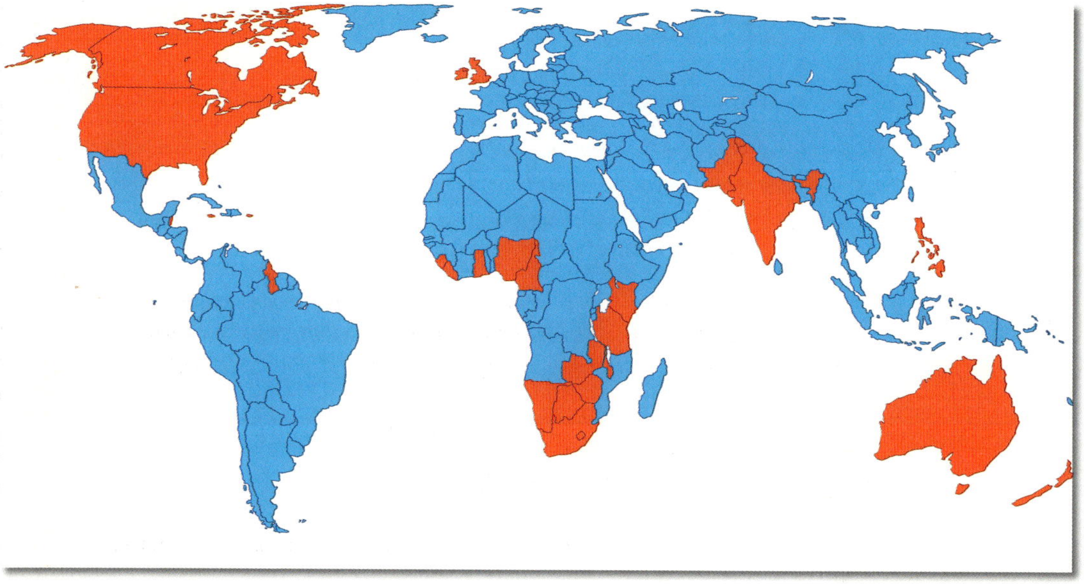

Work with a partner. Make the following lists:

1 Countries where English is an official language. (Refer to the map.)
2 English-speaking countries in which you would like to work. Explain why.
3 The aspects of your job for which you need English.

2 APPLYING FOR A JOB ABROAD

Read the advertisement on the opposite page, then answer the questions.

1 Where is the job?
2 What is the job title?
3 What information are we given about the company?
4 What sort of person can apply for this job?
5 How can you earn more money than your normal salary in this job?
6 How do you know the job is not for a short time only?
7 How can you apply for this job?
8 What document must you send?

Back Forward Stop Refresh Home AutoFill Print Mail

Address: http://www.techjobsinternational.com

SEARCH BY LOCATION | BROWSE JOBS | REGISTER YOUR CV | COMPANIES A-Z | HELP

JOB DETAILS

Tekskil Recruitment Agency

BEGIN A NEW SEARCH

Switchboard/Panel Wirer (permanent position)

Location: Sydney, New South Wales
Country: Australia
Salary: negotiable
Reference: SBPW2000
Date posted: 19/04/20..

Our client is a manufacturer of industrial machinery. If you are a licensed electrician or electrical fitter, want a permanent job and want to work in one location then apply for this rewarding job.

Benefits include:

- a generous salary
- plenty of overtime at attractive rates
- job security

Call 0061 2 9637 8600 and ask for Lisa Mallory or apply now online:

You will be asked to upload your CV (.doc, .txt or .pdf files only)

APPLY FOR THIS JOB

3 WORKING WITH WORDS

Find the English words in the text.

1 Ort
2 Kunde
3 Einzelheiten
4 Personalagentur
5 Dauerstelle
6 lizensiert
7 Gehalt
8 Überstunden
9 Sicherheit
10 hochladen
11 Lebenslauf
12 Datei

4 A CV IN ENGLISH

Philipp Wahler is a trainee electrician in Schweinfurt. He decides to apply for the job on page 117. He prepares his CV.

Work with a partner. Look at Philipp's CV on the opposite page. How is it different to a CV in German?

1 What isn't included?
2 What extra information does it contain?
3 What is in a different order?

5 SKILLS SUMMARY

As part of his application, Philipp has to write a 75–80 word summary of his skills.

Use the information in Philipp's CV (Work Experience, Relevant Skills, Extra-curricular Activities) to write the summary.

Example: *In the course of my 3-year traineeship in Germany I have gained experience of …*

CULTURE

Curriculum vitae (CV)

Im englischsprachigen Raum ist der Lebenslauf der wichtigste Bestandteil einer Bewerbung. Im amerikanischen Englisch heißt Lebenslauf *résumé*. Manchmal wird ein Begleitschreiben gar nicht verlangt. Der Lebenslauf sollte maximal zwei DINA4-Seiten (Schriftgröße 12) einnehmen.

INFORMATION

Tipps für den Lebenslauf

Die Erläuterungen beziehen sich auf die hochgestellten Ziffern im Lebenslauf auf Seite 119.

1 Schreiben Sie den Monat immer voll aus: *May 2006* statt *06.2006*, usw.
2 Das Pluszeichen ersetzt den Auslandscode 00.
3 Hier erfährt der Arbeitgeber auf einen Blick, wer Sie sind und was Sie wollen.
4 Weitere Berufsbezeichnungen: *electronic technician* – Elektroniker/in; *energy electronics technician* – Energieelektroniker/in; *communications electronics technician* – Kommunikationselektroniker/in.
5 Die Beherrschung einer Fremdsprache wird mit *fluent*, *good* und *basic* beschrieben.
6 Alle deutschen Begriffe (einschl. Firmennamen) sind ins Englische zu übersetzen.
7 Eine knappe Bezeichnung Ihrer schulischen Laufbahn reicht.
8 Alle weiteren für das Arbeiten im Ausland relevanten Fähigkeiten sind hier zu erwähnen.
9 Geben Sie hier außerschulische Aktivitäten an, die Aufschluss über Ihre Persönlichkeit geben.
10 E-Mail-Adressen erleichtern die Einholung der Referenzen.

CURRICULUM VITAE

PERSONAL DETAILS

Name:	Philipp Wahler
Date of birth:	06 April[1] 1987
Gender	Male
Nationality:	German
Marital status:	Single
Address:	Siedlerweg 26, 97422 Schweinfurt, Germany
Telephone:	+49 9721 79664[2]
Mobile phone:	+49 179 4327791
Fax:	+49 9721 79665
Email:	philipp.wahler@gmx.de

PROFILE[3]

Trainee electrician[4] in final year of 3-year course seeking post abroad.
Good[5] spoken and written English. Mobile and flexible.

EDUCATION AND TRAINING

Since 2003	3-year traineeship at Buderus Medical Care AG[6] and Dr Georg Schäfer Vocational Training College in Schweinfurt, Germany
June 2003	German Class 10 School Leaving Certificate
1993–2003	Primary and secondary education in Schweinfurt[7]

WORK EXPERIENCE

Since 2003	Experience of installing, maintaining and repairing electrical systems (household and industrial wiring) and motors
2002	5-day placement at Elektro Wilhelm (local electrical contractor) doing maintenance and repairs of household wiring 3-day placement at SKF Gebäudetechnik GmbH (energy-efficient building technology)

RELEVANT SKILLS[8]

Knowledge of CAD and technical drawing
Familiar with word-processing and spreadsheet programs
Full driver's license
First aid certificate
Life-saving certificate

EXTRA-CURRICULAR ACTIVITIES[9]

Voluntary member of local fire and rescue service
Captain of local soccer team
Active member of church community service group

REFEREES

Herr Thomas Weinberg
Training Officer
Buderus Medical Care AG
Gustav-Adolf-Straße 8
97422 Schweinfurt
Germany
Tel. +49 9721 7193 35 56
Email: thomas.weinberg@buderusMC.de[10]

Frau Jutta König
Class Teacher
Dr. Georg-Schäfer-Schule
Geschwister-Scholl-Straße 18
97424 Schweinfurt
Germany
Tel. +49 9721 93 99
Email: jutta.könig@bs1-sw.de

6 CV HEADINGS

Match the information from a CV (a–l) with the headings (1–6). There are two facts for each heading.

1 Profile
2 Education and Training
3 Work Experience
4 Relevant Skills
5 Extra-curricular Activities
6 Time Spent Abroad

a Travelled widely in southern Africa
b Regular holiday jobs as electrician's mate
c Licensed electrician with experience of work in Arab countries
d Webmaster for website of local rock band
e 3 months spent at Orlando High School, Florida, USA at the age of 16 as an exchange student
f Member of church choir
g Primary school in Romania
h Construction site work in commercial premises and private houses
i Trainee electronic technician with basic English
j Fluent spoken and written Turkish
k June 20.. German higher education entrance qualification
l Goods vehicle licence

7 YOUR STRONG POINTS

Make a list of your own skills and strong points under the headings in exercise 6. You can invent more headings if necessary.

8 EDUCATION, TRAINING AND SKILLS

Find the English equivalent (a–j) for each German expression (1–10).

1 Allgemeine Hochschulreife (Abitur)
2 Ausbildung, Lehre
3 Azubi, Lehrling
4 Berufsschule, Berufskolleg, Gewerbliche Schule
5 Fachhochschulreife
6 Führerschein Klasse 3
7 Mittlere Reife
8 Sekundarstufe 1 und 2
9 Wehrdienst
10 Zivildienst

a secondary education
b military service
c traineeship, apprenticeship
d German school leaving certificate (year 10)
e community service
f vocational college (GB), vocational school (US)
g trainee, apprentice
h category B (car) driving licence
i German university entrance qualification
j German higher education entrance qualification

9 PHILIPP'S COVER LETTER

Read Philipp's cover letter and find the following information. Write down the line number(s).

1 When he can start
2 His skills
3 His training
4 What job he is applying for
5 What he is sending with the letter

Siedlerweg 26
97422 Schweinfurt
Germany
Tel. +49 9721 79664
Mobile: +49 179 4327791
Email: philipp.wahler@gmx.de

7 May 20..

Ms Lisa Mallory
Tekskil Recruitment Group
180 Market Street
Merrylands NSW 2160
Australia

Dear Ms Mallory

Vacancy for Switchboard / Panel Wirer dated 19 April 20.., Ref. SBPW2000

With reference to your advertisement on the internet, I wish to apply for the above-mentioned post.

By June of this year I will have completed a three-year apprenticeship at Buderus Medical Care AG with theoretical instruction at Dr Georg Schäfer Vocational Training College in Schweinfurt, Germany. I will then be a licensed electrician.

In the course of my apprenticeship I have gained experience of household and industrial electrical installation, including all the usual safety features and also maintenance and repair work. I have learned to work reliably and independently in environments ranging from hospital operating theatres to building sites.

My CV is enclosed for your information and I am available for interview at short notice.

As I will shortly have the qualifications you specify together with a good command of spoken and written English, I would be grateful if you would consider me for this post.

Yours sincerely

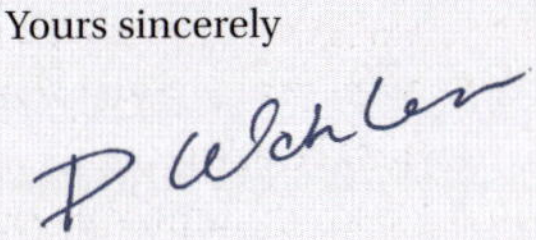

Philipp Wahler

Enc

10 TIPPS FÜR BEWERBUNGSSCHREIBEN AUF ENGLISCH

Schauen Sie sich das Bewerbungsschreiben auf Seite 121 nochmals an. Schreiben Sie diese Hinweise ab und ergänzen Sie die Lücken.

1 Schreiben Sie Ihre Adresse … auf dem Blatt.
2 Vergessen Sie nicht, sämtliche … anzugeben.
3 Die Betreffzeile kommt … der Anrede und wird … .
4 Wenn Sie eine bestimmte Person anschreiben, lautet die Schlussformel: *Yours …*
5 Der Name wird unter der … ausgeschrieben.
6 Wenn Sie eine Anlage (z.B. Ihren Lebenslauf) mit dem Brief schicken, schreiben Sie unten: … .

11 APPLYING FOR A JOB

You find this job advertisement on a website. Apply for the job:

1 Write your CV.
2 Write a 75–80 word summary of your skills.
3 Write a cover letter.

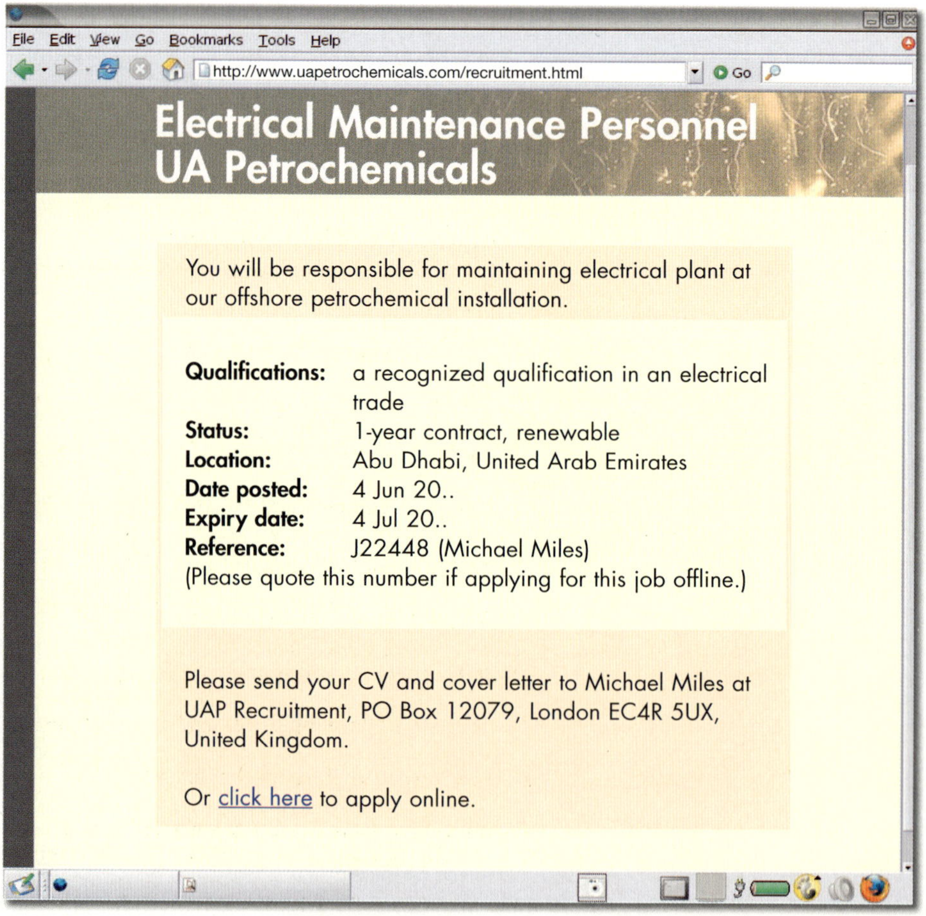

12 JOB INTERVIEW

23 Maik Gärtner applies for the Abu Dhabi job and is invited to London for an interview.

a These are some of Maik's answers in the interview. Which question (1–10) from the magazine article does each answer go with?

a I like building computers and robots.
b It's in the family. My dad is an electrician.
c At the moment I'm just finishing a three-year training course.
d I want experience in a big company outside Germany.
e I enjoy working with other people and I play in a football team.
f How old are the other employees in Abu Dhabi?
g I don't have much experience.
h I want to set up my own business.
i I expect to get a good final grade at the end of my training.
j I'm hard-working, punctual and reliable.

b Now listen to the CD to check your answers.

TOP TEN
INTERVIEW QUESTIONS

We asked 100 personnel managers:
'What are your top ten questions for recruitment interviews?'
Here are the ten most popular questions:

1 Please take us through your CV.
2 Are you a team player?
3 Do you have any hobbies?
4 What are your strengths?
5 What are your weaknesses?
6 Why did you decide to become a … ?
7 Why do you want this job?
8 Why should we employ you and not someone else?
9 What are your plans for the future?
10 Do you have any questions you'd like to ask?

13 CHECKLIST

23

Listen again to the interview and complete the checklist with notes about Maik.

Candidate: Maik Gärtner

1 Is the candidate able to talk about him/herself? Yes. Fluent description of his course and experience.

2 What sort of person is the candidate? Works for Federal Agency for Technical Relief (THW). Knows what to do in an emergency.

3 Is the candidate a team player? Example(s)?

4 What are the candidate's hobbies?

5 What are the candidate's strengths and weaknesses?

6 What are his/her reasons for becoming an electrician?

7 Why does the candidate want this job?

8 Why should we employ this particular candidate?

9 What are the candidate's plans for the next 5 to 10 years?

10 Candidate's questions to us:

14 JOB INTERVIEWS

Work in groups of three.

Student A: Turn to file 8 on page 128.
Student B: Turn to file 9 on page 129.
Student C: Turn to file 17 on page 132.

GRAMMAR

-ing forms and infinitives

1 I **enjoy working** with other people.
2 I **expect to get** a good final grade.
3 I **want to set up** my own business.
4 I **like building** / **to build** computers and robots.
5 You will be responsible **for maintaining** electrical plant.

- Die *-ing*-Form folgt auf Verben wie *enjoy, involve, mind, stop, suggest.* (1)
- Auf andere Verben wie *agree, choose, decide, expect, hope, learn, want* folgt der Infinitiv. (2, 3)
- Auf einige Verben wie *begin, continue, hate, like, love* und *start* kann sowohl die *-ing*-Form als auch der Infinitiv folgen. (4)
- Wenn auf eine Präposition ein Verb folgt, muss man die *-ing*-Form verwenden. (5)

15 PRACTICE

Complete the sentences with the correct form(s) of the verbs in brackets.

1 I don't mind … (do) overtime.
2 Why did you choose … (become) an electrician?
3 Maik hopes … (start) his own company one day.
4 I don't think he has much hope of … (get) this position.
5 She soon began … (get) bored with the job.
6 The company hasn't agreed … (take) Maik on yet.
7 The successful candidate will be expected … (work) shifts.
8 Do you think Maik is interested in … (learn) Arabic?
9 He doesn't really enjoy … (work) as part of a team, I think.
10 I'd really hate … (live) in a hot country.

EXTRA MATERIAL

DOs and DON'Ts for JOB INTERVIEWS

As you come into the room, you have a price tag hanging around your neck and the employer is picking you up, shaking you, looking at you from all angles wondering, 'Is he/she really worth that much?'

• DO

Look professional – not trendy. Wear neutral tones (darker is more impressive), nothing loud or shiny.
Make sure your clothes are perfectly clean. They needn't be expensive but they must be in perfect condition.
Wear a good-quality shirt/blouse in a colour that suits your skin.
Wear sensible shoes.
Men: wear good socks. (You can wear your Mickey Mouse socks later, when you've got the job.)
Women: choose light-coloured tights. (Dark tights are evening wear.)
Men: shave well in advance so your skin has time to settle.
Women: wear only a little make-up to bring out your best features.

• DON'T

Don't put your hands in your pockets.
Men: don't have long hair – get it cut a week before your interview.
Men: don't use strong aftershave – the interviewer may hate it!
Women: don't use perfume (see above).
Don't bring your mobile phone with you, or at the very least, turn it off.
Don't wear lots of jewellery. This is seen as a sign of insecurity. Don't wear that chunky gold necklace or bracelet you wear at the disco. Wear a maximum of one ring on each hand.
Men: don't wear big chunky watches. You are not James Bond.

• ON THE GREAT DAY

Telephone to confirm that you're coming. Do this the day before if the interview is in the morning.
Don't eat strong-smelling foods before the interview. You can smell garlic for at least 24 hours.
Arrive five to ten minutes early. Allow plenty of time for traffic and parking. Sit in a nearby car park or café with a book or newspaper.
Give your coat (and umbrella) in at reception.
Ask to use the bathroom and then: check your hair; women: check your make-up; clean your glasses; re-polish your shoes; wash your hands in warm water; dry them thoroughly; check your nails are clean; check your clothing a final time.

Adapted from: Rowan Manahan, *Where's My Oasis?* Random House, 2004

1 UNDERSTANDING THE TEXT

According to the text, what is the candidate in the cartoon doing wrong?

I'll be with you in a moment.

2 DISCUSSION

Work with a partner or in a small group.

1 Choose the two DOs and DON'Ts which are most important in your opinion. Explain your choice in a few sentences.
2 Are there any tips here that you disagree with? Explain why.

Appendix

PAIRWORK FILES

File 1 UNIT 2, EXERCISE 12

Student A

1 You need a new multimeter. (Use language from exercise 11.)

- Phone TechTools and ask for an M3900 digital multimeter.
- If the multimeter isn't in stock, ask what other multimeters they have.
- Ask the price.
- Ask whether TechTools can invoice your company.

2 You are the salesperson at TechTools. Take the customer's order. (Use language from exercise 11.)

MK6228	Makita cordless drill 14.4 V	€135.95	0 in stock
32614-2G	Bosch cordless drill 15.6 V	€148.99	3 in stock

Payment terms on cordless drills: credit card only.

File 2 UNIT 5, EXERCISE 14

Student A

Sie sind Projektleiter/in in einer Vernetzungsfirma, DataNet. Sie beginnen das Gespräch.

1 Stellen Sie sich vor und fragen Sie, womit Sie helfen können.
2 Fragen Sie, wie viele Mitarbeiter (*employees/staff*) die Firma hat und wie viele Computer vorhanden sind. Fragen Sie den Kunden nach seinen Bedürfnissen (*requirements*).
3 Schlagen Sie eine Sterntopologie vor. Verwenden Sie dabei Ausdrücke aus dem *Language*-Kasten (S. 50).
4 Beantworten Sie die Fragen des Kunden.
5 Wenn der Kunde darum bittet, versprechen Sie, einen Kostenvoranschlag (*quotation*) zu schicken. Bitten Sie den Kunden um seine Adresse, Telefon- und Faxnummer und E-Mail-Adresse.
6 Verabschieden Sie sich.

File 3 UNIT 6, EXERCISE 7

Student A

1 Read these sums to your partner. Check that he/she has written them down correctly.

1 $12 \times 12 \times 12 = 12^3$
2 $246 - 83 = 163$
3 $4{,}698 \times 67 = 314{,}766$
4 $0.4 \div 0.08 = 5$
5 $5 + (60 \div 12) = 10$
6 2 kilovolts = 2,000 volts
7 1 gigawatt = 1,000,000,000 watts
8 ¼ + ½ = ¾

2 Write down the sums that your partner reads to you.

File 4 UNIT 6, EXERCISE 13

Student A

1 Describe to your partner how to install a stand-alone PV system in a house. The expressions in the box will help you, but you have to put them into the correct order.

connect inverter ■ connect charge controller
connect PV modules in series ■ install PV array on roof
connect to domestic wiring system ■ connect batteries

2 Listen to your partner's explanation. Does it sound correct to you? If not, ask questions to check.

File 5 UNIT 8, EXERCISE 2

Student A

1 Write down the values that your partner reads to you, then ask your partner to check your answers.

2 Read out these values to your partner.

1 90 µs
2 23,000 V
3 150 µF
4 10,000 W
5 50.5 Hz

File 6 UNIT 8, EXERCISE 4

Student A

1 **Ask your partner the following problems. Then check your partner's answers against the answers provided.**

Begin like this:
What's the root mean square / peak voltage if ...?

1	PV = 50 V	(Answer: rms = 35.361 V)
2	rms = 308.5 V	(Answer: PV = 436.219 V)
3	PV = 2,000 V	(Answer: rms = 1,414.427 V)
4	rms = 268,741.15 V	(Answer: PV = 379,999.98 V ≈ 380,000 V)

2 **Write down the problems that your partner reads out.**
Use a calculator to work out the answers ($\sqrt{2} \approx 1.414$).
Read out the answers to your partner.

File 7 UNIT 9, EXERCISE 7

Student A

1 **You are a German domestic appliance technician.**
You have been called out to repair a customer's washing machine, but you do not have an up-to-date manual for that model. You phone the American manufacturer to ask for help.

Say what the problem is, then answer the American technician's questions to find out the likely cause.

Problem:	The machine does not fill properly.
Indicators:	LED X5 is off.
	LED Y27 is on.
	LED Y12 is on.
Other information:	The drain is open.

2 **You are an American washing machine technician. You receive a call from a German colleague who needs your help. Use the troubleshooting guide on page 89 to ask relevant questions and diagnose the fault.**

Example: *Is LED Y0 on or off?*
Have you checked the ... ?
It sounds as if the fault is probably ...

File 8 UNIT 12, EXERCISE 14

Student A

You are an interviewer. You are interviewing candidates for jobs in the construction industry in various parts of the USA.
You don't know much about electrical engineering, but you want to find out what kind of person the candidate is.

- Ask the candidate why he/she wants this job.
- Ask the candidate about his/her hobbies.
- Ask the candidate about his/her strengths and weaknesses.
- At the end of the interview, ask the candidate whether he/she has any questions.

File 9 UNIT 12, EXERCISE 14

Student B

You are an interviewer. You are interviewing candidates for jobs in the construction industry in various parts of the USA.
You are an electrical engineering expert, and you want to find out whether the candidate has the right technical skills.

- Ask the candidate about his/her reasons for becoming an electrician.
- Ask about his/her training.
- Ask about his/her experience so far, e.g. work placements.
- Ask whether he/she has any other relevant skills.

File 10 UNIT 2, EXERCISE 12

Student B

1 You are the salesperson at TechTools. Take the customer's order. (Use language from exercise 11.)

M3900	Digital multimeter	€39.95	0 in stock
VC890D	Digital multimeter	€45.00	12 in stock

Payment terms on multimeters: credit card or cash on delivery only.

2 You need a new electric drill. (Use language from exercise 11.)

- Phone TechTools and ask for a Makita 14.4V cordless drill.
- If the drill isn't in stock, ask what other cordless drills they have.
- Ask the price.
- Ask whether TechTools can invoice your company.

File 11 UNIT 5, EXERCISE 14

Student B

Sie sind der Kunde / die Kundin. Bereiten Sie sich auf das Gespräch vor, indem Sie die Informationen im Kästchen lesen.

> Sie haben eine kleine Firma, EcoGreen, die umweltfreundliche Reinigungsmittel (*eco-friendly cleaning products*) verkauft.
> Ihre Firma hat zwei Partner, drei Kaufleute und einen Buchhalter (*bookkeeper*).
> Kundendaten werden auf unvernetzten Computern gespeichert. Sie befürchten, sie könnten bei einem Absturz verloren gehen.

Ihr/e Partner/in wird das Gespräch beginnen.

1 Stellen Sie sich vor und erklären Sie, worum es geht.
2 Beantworten Sie die Fragen des Netzwerkexperten.
3 Hören Sie sich die Vorschläge des Experten an und erkundigen Sie sich nach den Vorteilen dieses Netzwerks. Verwenden Sie dabei Ausdrücke aus dem *Language*-Kasten (S. 50)
4 Akzeptieren Sie schließlich den Vorschlag. Bitten Sie um einen Kostenvoranschlag (*quotation*).
5 Geben Sie Ihre Adresse, Telefon- und Faxnummer und E-Mail-Adresse an, wenn der Experte darum bittet.
6 Verabschieden Sie sich.

File 12 UNIT 6, EXERCISE 7

Student B

1 Write down the sums that your partner reads to you.

2 Read these sums to your partner. Check that he/she has written them down correctly.

1 $2 \times 2 \times 2 \times 2 = 2^4$
2 $324 - 78 = 246$
3 $9{,}842 \times 55 = 541{,}310$
4 $0.65 \div 5 = 0.13$
5 $10 + (6 \div 2) = 13$
6 21 megaohms = 21,000,000 ohms
7 1 terabit = 1,000,000,000,000 bits
8 $\frac{1}{2} - \frac{1}{3} = \frac{1}{6}$

File 13 UNIT 6, EXERCISE 13

Student B

1 Listen to your partner's explanation. Does it sound correct to you? If not, ask questions to check.
2 Describe to your partner how to install a grid-tied PV system in a house. The notes below will help you but you have to put them into the correct order.

connect array with inverter ■ install electric meter
connect PV modules in series ■ install PV array on roof
connect system to national grid

File 14 UNIT 8, EXERCISE 2

Student B

1 Read out these values to your partner.

1 2.3 Ω
2 90 ms
3 0.5 A
4 5 μH
5 45,000 kWh

2 Write down the values that your partner reads to you, then ask your partner to check your answers.

File 15 UNIT 8, EXERCISE 4

Student B

1 Write down the problems that your partner reads out.
Use a calculator to work out the answers ($\sqrt{2} \approx 1.414$).
Read out the answers to your partner.

2 Ask your partner the following problems. Then check your partner's answers against the answers provided.

Begin like this:
What's the root mean square / peak voltage if …?

1	PV = 75 V	(Answer: rms = 53.041 V)
2	rms = 230 V	(Answer: PV = 325.22 V)
3	PV = 25,000 V	(Answer: rms = 17,680.339 V)
4	rms = 353,606.78 V	(Answer: PV = 499,999.98 V ≈ 500,000 V)

File 16 UNIT 9, EXERCISE 7

Student B

1 You are an American washing machine technician. You receive a call from a German colleague who needs your help. Use the troubleshooting guide on page 89 to ask relevant questions and diagnose the fault.

Example: *Is LED X5 on or off?*
Have you checked the … ?
It sounds as if the fault is probably …

2 You are a German domestic appliance technician.
You have been called out to repair a customer's washing machine, but you do not have an up-to-date manual for that model. You phone the American manufacturer to ask for help.

Say what the problem is, then answer the American technician's questions to find out the likely cause.

Problem:	The wash motor will not start.
Indicators:	LED Y0 is on. LED Y4 is on. LED Y5 is off. LED Y1 is off.
Other information:	The motor starter and heater coils are fine. The brake seems to be off and the door switch circuit seems OK.

File 17 UNIT 12, EXERCISE 14

Student C
You are the candidate. You have answered an advertisement for electricians to work in the construction industry in the USA. You love to travel and you have always wanted to visit the USA.

Read the information below before the interview.

Profile:	You are an independent person: you always get the job done and you don't need a lot of supervision. (Your weakness: sometimes you are impatient.)
Training:	You have just finished a three-year training course at Elektro Kaiser (rewiring old houses and repairing heating systems) and a local vocational college.
Hobbies:	You like repairing old radios, tape-recorders, etc. You are a member of a canoeing club. You help to train young children to canoe.
Work experience:	Last summer you had a holiday job on a construction site helping electricians (mainly fetching tools and cleaning up).
Questions:	You want to know if there is a chance of a job in California.

TRANSCRIPTS

3 UNIT 1, EXERCISE 13: A GUIDED TOUR OF FISCHER ELEKTRONIK

JOCHEN Here we are. This is the visitors' car park.

EDDIE How many people do you employ here?

JOCHEN We employ 51 people here, including three master electricians, 12 qualified electricians and six trainees.
(Jochen parks the car.)

JOCHEN Now let's go in through reception in the main administration building here in front of us …
(They go into the building.)

JOCHEN The canteen's just next door here on our left – would you like a cup of coffee before I show you around?

EDDIE No, thank you very much. I've got another appointment in an hour or so.

JOCHEN Fine. Let's just walk through reception past the stairs here and turn left into the main corridor, which links all the departments. Now here, on your right, is our solar technology section. We have two technicians here. Let's have a look inside.
(They go in.)

EDDIE Is there much demand for solar energy?

JOCHEN More and more, as energy prices rise.

EDDIE I see. *(They go back outside.)* What's this department next door to solar technology?

JOCHEN That's lighting technology. We do everything from a restaurant to a football field. Opposite lighting technology we have electrical installation. Let's have a look inside, shall we? *(They go in.)*

EDDIE There aren't many people here.

JOCHEN No, that's because most of them are out on jobs. At the moment the two trainees you can see are working on a project for college.

EDDIE I see.

JOCHEN Let's take a look at our communications technology section, shall we? It's the next door on the left.

EDDIE Is this mainly telephones?

JOCHEN Most of our work has to do with phone systems but we also install satellite dishes and digital TV and radio aerials.

EDDIE How about computers?

JOCHEN Our network technology section is the next door on the left at the end of the corridor. It's a big field nowadays. Would you like to take a look at our equipment?

EDDIE No, thanks, Jochen. I'm a businessman, so I'll leave the complicated stuff to your experts!

JOCHEN Well, we've almost finished our tour now. On the other side of the corridor you can see our after-sales department, which provides service and support for our customers.

EDDIE Yes, that's very important. How many technicians do you have?

JOCHEN We have six maintenance technicians, each with a fully equipped van. They can handle most problems on the spot. Let's walk through after-sales to the last section – the warehouse.
(*They go into the warehouse.*)

EDDIE It's pretty big and well-stocked. How many different articles have you got here?

JOCHEN We stock about 12,000 articles, so the customer doesn't need to wait long for service. The big double doors you can see here on your left lead to our other car park for maintenance and delivery vehicles. Now is there anything else you'd like to know?

5 UNIT 2, EXERCISE 11: ORDERING TOOLS

TECHTOOLS TechTools, Lee speaking. How can I help you?
BEN Hello, I'd like to place an order.
TECHTOOLS Of course. What would you like, sir?
BEN I'd like to order a 300 mm vernier calliper, please.
TECHTOOLS I'm afraid we don't have any 300 mm vernier callipers in stock.
BEN Do you have any other vernier callipers?
TECHTOOLS Yes, we have a 200 mm vernier calliper.
BEN How much is that?
TECHTOOLS It's €48.95.
BEN I'll have one of those, please.
TECHTOOLS How would you like to pay?
BEN Can you invoice my company?
TECHTOOLS I'm sorry. We only offer credit card payments or cash on delivery to new customers.
BEN That's OK. Cash on delivery, then, please.

8 UNIT 3, EXERCISE 9: NEW WIRING FOR THE KITCHEN

PENNY Well, there aren't many fittings here at the moment because the other people had a gas cooker. There's just a ceiling light, a light switch by the door here on the left and a single socket over there on the left half-way along the wall.
FRANK Yes, I see. I've marked that. Now what else do you need?
PENNY Well, first I need another socket over the light switch and another one on the left here on the same wall in the corner.
FRANK (*Writing it down*) … On the left … on the same wall … in the corner … OK, I've got that. Now where do you want the cooker?
PENNY On the left-hand wall.
FRANK At the far end, near the window?
PENNY No, at this end, near the door.
FRANK Right, so we'll need a cooker socket there with five-core cable.
PENNY And I'll need another ceiling light in the middle, here near the door, too.
FRANK (*Writing it down*) … A ceiling light in the middle, near the door … Just a minute! You'll have a light over the cooker with the extractor fan anyway.
PENNY Oh, yes. I was forgetting.
FRANK So, you'll need a socket for the extractor fan, too. Let's say to the right of the cooker, OK?

PENNY All right, so we don't need the second ceiling light.

FRANK Good. Now what else do you need?

PENNY We'll need a socket just here on the right-hand wall for the fridge-freezer combination.

FRANK Right … I've got that.

PENNY The dishwasher and sink unit are going to be in the far corner on the left, so we'll need a socket on the left-hand wall near the corner and another ceiling light over the sink to the left of the dishwasher. Oh yes, and as there's only a single socket on that wall at the moment, I want you to change that into a triple socket.

FRANK OK. Any more sockets? How about a double socket in the far corner on the right, for example, or between the two windows?

PENNY Oh, I think a single one will do there.

FRANK (*Writing it down*) … A single between the windows.

PENNY No, I mean a single in the far right-hand corner. We don't need a socket between the windows.

FRANK Fine … Have we got everything now?

PENNY Err, yes. I think so.

FRANK Great. Well, I must rush. Goodbye!

PENNY Bye!

11 UNIT 5, EXERCISE 11: ADVISING THE CUSTOMER

PETER Thanks for your email, it was very useful. Now, how do you want your PCs to be connected to your LAN?

ROBERT Sorry, what's a 'LAN', exactly? I don't know much about computers.

PETER A LAN is a 'local area network', so, basically, your network in your school.

ROBERT Ah, right … Could you give me some information about the different possibilities, please?

PETER Sure. Basically there are three topologies … that means arrangements of a network. In the bus topology each workstation or device is connected to a main coaxial cable called a bus.

ROBERT A bus, right …

PETER Now, a bus network is simple and cheap.

ROBERT Sounds good …

PETER On the other hand, bus networks can't support heavy traffic. If you are all using your computers at the same time, the network will slow down.

ROBERT Oh, I see … I don't think that would work for us, then.

PETER No, I agree. Another possibility is a ring topology. So called because all the devices are connected to each other in a ring. Data is passed around the ring from one device to the next until it reaches the device which needs it.

ROBERT OK. I think I've got that.

PETER The good thing about ring topologies is that they can support heavy traffic. Even when all the people in your office use the network at the same time, it will work fast and without any problems.

ROBERT I see. That sounds like a good idea.

PETER Yes, but on the other hand, ring topologies are expensive. I don't think it's the best option for you.

ROBERT Oh … Well, what would you suggest, then?

PETER I'd recommend a star topology. It's so called because all the devices are connected by separate cables to a central unit, called the switch.

ROBERT And that's good, is it? Sounds like a lot of cables …

PETER True, but star networks are easier to manage than bus and ring networks. It's also easy to add new workstations.

ROBERT Right, so as our school grows, we can add more computers?

PETER That's right. But there is one disadvantage. If the switch is faulty, the system won't work. But that doesn't happen often.

ROBERT OK … Well, that sounds fine. Let's go with a star topology, then.

PETER Good. Now, here's a plan I've drawn up …

13 UNIT 6, EXERCISE 10: GRID-TIED OR STAND-ALONE?

MAREN So, we're agreed that I want a PV system, then. It will be good to be independent of the national grid.

CHRISTIAN Well, that's not entirely correct. You're talking about a stand-alone PV system. That wouldn't be connected to the national grid. That's why stand-alone systems are used where there's no access to mains electricity.

MAREN Right … I think I see …

CHRISTIAN Take a look at this diagram, it will show you how the stand-alone system works. A PV array on the roof of the house generates direct current, which is stored in batteries. A charge controller maintains the batteries at the proper charge level and protects them from overcharging.

MAREN So everything in the house has to run on direct current?

CHRISTIAN No. An inverter changes the direct current to alternating current, because most standard appliances only run on alternating current.

MAREN OK. Is that the PV system I need?

CHRISTIAN Not really. Your home is already connected to the national grid, so I'd recommend a grid-tied PV system.

MAREN Why's that?

CHRISTIAN It's much less expensive to buy and operate. You don't need batteries or a charge controller – as you can see in this diagram here.

MAREN Oh, I see. How does it work, then?

CHRISTIAN Your PV array produces direct current in the normal way, and an inverter changes it into alternating current. Then the electricity you produce is fed straight into the national grid.

MAREN So how does that help me?

CHRISTIAN Your electricity provider buys the electricity from you. Then you take electricity from the national grid in the normal way for all your household appliances.

MAREN Right. Is that why there are two electricity meters in the diagram?

CHRISTIAN That's right. The power you sell to the grid is measured by one, and the power you buy from the grid is measured by the other.

MAREN OK … Do I need to get my electricity company to install the meters?

CHRISTIAN No, you don't. The meters will be installed by us as part of our service.

14 UNIT 7, EXERCISE 6: GETTING A SECOND OPINION

ARMIN This is Kilo Sierra nine Oscar Mike – KS9OM – in Freiburg, Germany.
CARLA This is Alpha Alpha four Mike India – AA4MI – in Portland, Oregon. Hi, how are you doin'?
ARMIN Good. How about you?
CARLA I'm doin' fine.
ARMIN Can you help me repair a guitar amp, Carla?
CARLA I'll do my best, Armin. What's wrong with it?
ARMIN Well, it doesn't work at all. A fuse was blown, so I've replaced it but it still doesn't work. I've had a good look at it and there's no smell of burning or charred wires and the valves are illuminated,
CARLA I hope your shack has a GFCI, Armin. If not you could electrocute yourself!
ARMIN A GFCI? That's a 'ground fault circuit interrupter', right?
CARLA That's right.
ARMIN We call that type of circuit breaker an RCD over here, Carla. Don't worry: I have one in the circuit.
CARLA That's good to hear. Do you have the schematic of the amp?
ARMIN Yes, I found it inside the case of the amp.
CARLA Have you started tracing the signal?
ARMIN Yes, I have, using a digital multimeter. I haven't got very far yet but I think I've found something. The voltage input is 230 volts but one of the five output voltages from the transformer is zero. It should be 430 volts.
CARLA OK. The transformer could be faulty, or there could be a short circuit somewhere.
ARMIN Do you think I should replace the transformer?
CARLA Not yet. It may be OK. What is attached to it where the voltage is zero?
ARMIN A full-wave bridge rectifier.
CARLA Then disconnect that and measure the transformer output voltage to see if it's still zero. If the transformer's OK, the fault is somewhere else.
ARMIN OK, I'll do that. What time is it over there? It's midday here.
CARLA It's 3 a.m.
ARMIN Oh, no! Sorry to bother you so late.
CARLA No problem – this is the best time for radio hams! Good luck with the amp!

16 UNIT 8, EXERCISE 7: MAKING A BATTERY CHARGER

MATTI Could you explain something to me, Dennis? I think I've understood but I'm not quite sure. Does Herr Böhm want us to make what it says in this circuit diagram?
DENNIS Yes, that's right. It's a battery charger. We have to get all the components we need, assemble them on a circuit board and test the battery charger at different points and explain how it works.
MATTI I see. Will we have to buy the parts?
DENNIS No, we won't. We have them all here in our store.
MATTI Oh, I see. I'm not quite sure what sort of battery we will be able to charge with it. Will it be a car battery?
DENNIS No, look here ... The voltage regulator in the circuit only supplies a low amperage and you need about 10 amps for a car battery.

MATTI I see. But there's an adjustable voltage regulator in the circuit. Does that mean it can recharge different types of battery?

DENNIS Yes, but only a low-power battery. This type of circuit can't charge a car battery.

MATTI And do we have to test the battery charger on a rechargeable battery?

DENNIS We can if we want, of course, but we don't have to. Herr Böhm's more interested in the readings we get when we test the circuit. Let's go to the store with a list of all the parts we need and then look up the voltage regulator on the internet.

17 UNIT 8, EXERCISE 9: CHECKING THE COMPONENTS

MATTI I think I've got all the components we need now.

DENNIS Good, but we'd better check to make sure. We mustn't forget anything, otherwise it won't work! You tell me what we've got and I'll make sure we've got it.

MATTI OK. We've got a 2 A fuse and a 0.2 A fuse and a …

DENNIS Just a moment – not too fast! They're pretty small … Right, we've got them. Now what's the next part?

MATTI A transformer with a 12 V output.

DENNIS We've certainly got the transformer. It's so big you can't miss it! How about diodes? How many of them have we got?

MATTI We've got five diodes all the same. They're 1N4004 diodes.

DENNIS One, two three four, five – OK, we've got them. Now have we got more than one of anything else?

MATTI Yes, we've got two electrolytic capacitors.

DENNIS Both the same?

MATTI No. One's a 2,200 µF / 25 V electrolytic capacitor.

DENNIS Just a minute. I'm just checking what it says on them. The writing's very small. What's the other one?

MATTI The other one's a 10 µF / 25 V electrolytic capacitor.

DENNIS Yes, we've got both. One's a lot bigger than the other one. Anything else?

MATTI We've got three more things on our list. There's the LM117 adjustable integrated voltage regulator …

DENNIS Here it is and it's got a heat sink to stop it from getting too hot.

MATTI Ah yes. I see. OK, the next component is a 220 Ω resistor.

DENNIS Another very small part but they're colour-coded to make it a bit easier to identify them. Right, we've got it. And the last part?

MATTI The last part is a 4.7 kΩ potentiometer.

DENNIS OK, we've got that, too. Now we can get started.

19 UNIT 10, EXERCISE 11: A PRESENTATION ON CULTURAL DIFFERENCES

SALLY Good morning, ladies and gentlemen. My name is Sally Bruce. Today, I'm going to talk about cross-cultural awareness. If you have any questions, I'd prefer to take them at the end.

To begin with, I'd like to point out that professionals working abroad should be prepared to find that everything is different from home. Becoming more aware of cultural differences will not only help you to communicate with others more effectively, but also to avoid misinterpretations. Important cultural patterns you should be aware of are attitudes towards time, business and company culture, body language and technical standards in a country. Let's illustrate this by looking at the two most important countries for our products: China and the USA.

In China being on time is vital. Always arrive on time or early if you are the guest. If you want to be successful in China, you have to understand and respect the Chinese business culture. Don't expect your Chinese business partners to make quick decisions: everything has to be discussed by the group.

As far as company culture is concerned, Chinese people value rank and status. The highest-ranking Chinese executive will be the one who does most of the talking, and your team should act in the same way.

Let's move on to body language. Chinese people don't usually make large hand movements while they are speaking. They don't touch each other either, except to shake hands when they meet.

Now it's time to talk about the USA. Again, being on time is essential. It is actually customary to be 10 to 15 minutes early. Business culture in the USA is the get-it-done style. American company culture tends to be individualist. Each executive has responsibility for his or her own area – they don't need to refer everything back to their boss.

Like the Chinese, American business partners don't touch each other much. In normal situations they stand about a metre apart: that's their personal 'comfort zone'. When greeting, Americans shake hands, and from an early age they are taught to do so with a firm, solid grip.

Technical standards are very important when doing business with the USA. Your products will need to conform to American standards as set by the National Institute of Standards and Technology.

Finally, remember that generalizing can be dangerous. The best strategy is to go to the other country and remember the two A's – be AWARE and ASK. Observe what people do and ask a lot of questions.

Well, if there are any questions, I'd be happy to answer them.

21 UNIT 11, EXERCISE 9: EN ROUTE TO GUANGZHOU

Dialogue 1

JÜRGEN Hi, I can't find the Air China check-in. I'm flying to Guangzhou via Beijing.

CLERK I'm sorry, sir. You're at the wrong terminal. You need Terminal 4.

JÜRGEN Oh, no! How do I get there?

CLERK Follow the signs to the shuttle train. The train leaves every ten minutes. When you arrive at Terminal 4, find Air China on the information board or ask an airport official to show you the way.

JÜRGEN Thank you.

Dialogue 2

CLERK Hello, sir. May I see your ticket and passport, please?

JÜRGEN Yes, certainly. Here you are.

CLERK Thank you ... How many pieces of luggage are you checking in, Mr Koller?

JÜRGEN Just this one, please.

CLERK Did you pack this bag yourself?
JÜRGEN Yes, I did.
CLERK And are you carrying anything for anyone else?
JÜRGEN No, I'm not.
CLERK And are you carrying any firearms, explosives or dangerous goods?
JÜRGEN No, I'm not.
CLERK That's fine, Mr Koller. I'll just print out your boarding pass.
JÜRGEN Err, could I have a window seat, please?
CLERK I'm sorry, sir. All the window seats are taken.
JÜRGEN Oh. Never mind.
CLERK Here's your ticket and boarding pass. Your flight is Air China flight number 538, departing at 17.40 to Beijing. Boarding time is 16.40, at gate 16.
JÜRGEN Thanks.
CLERK You're welcome. Have a good flight.

Dialogue 3

SECURITY Please put all your hand luggage on the belt … Is there a laptop in here, sir?
JÜRGEN Yes, there is.
SECURITY Could you take it out of the bag, please?
JÜRGEN Of course … Here you are.
SECURITY Thanks, sir. Please come this way.
(*The metal detector beeps.*)
SECURITY Do you have anything in your pockets?
JÜRGEN Oh, sorry. Yes, here are my car keys.
SECURITY Please put them in the tray and step through again, sir … OK, that's fine … Don't forget your keys.
JÜRGEN Oh, no. Thanks!

Dialogue 4

CUSTOMS May I see your passport and customs declaration card, please?
JÜRGEN Certainly.
CUSTOMS You haven't put your name on the card, sir.
JÜRGEN Oh, sorry! (*He writes his name.*) Here you are.
CUSTOMS Thank you. What's the purpose of your visit to China?
JÜRGEN It's a business trip. I'm going to install brewery technology for a plant here in Guangzhou.
CUSTOMS I see. Do you have anything to declare?
JÜRGEN Nothing except these bottles of our beer. I brought them along for my Chinese workmates. Am I allowed to bring them in?
CUSTOMS Visitors to China are allowed to carry two litres of alcoholic beverages, so that's no problem. Do you have any plants, vegetables or fresh foodstuffs?
JÜRGEN No, I don't.
CUSTOMS Here is your passport, sir. Give this customs declaration to the officer at the exit. Enjoy your stay in China.
JÜRGEN Thank you.

23 UNIT 12, EXERCISE 12: JOB INTERVIEW

JOHN Hello, Maik. My name's John Teesdale. Pleased to meet you.
MAIK Pleased to meet you.

JOHN Please take a seat.

MAIK Thanks.

JOHN Now I've got your CV here and I'd like you to take me through it.

MAIK Yes, certainly. At the moment I'm just finishing a three-year training course at an electrical contractor's called Elektro Wilhelm and it's based in Frankfurt. I've worked in private houses, shops, hospitals and on building sites doing electrical installation, maintenance and repair work … I like working on big projects, like factories or industrial sites.

JOHN Why's that?

MAIK It's good to work with other people, have a target and get the work done on time. I also like to work outside in the fresh air.

JOHN Right. Would you like to say a few words about your extra-curricular activities?

MAIK Yes, I'm a member of the Federal Agency for Technical Relief – what we call the THW – I'm in the Electric Supply group and I've learned a lot about what to do in an emergency.

JOHN I see. Would you describe yourself as a team player?

MAIK Yes, I'd say that. I'm in a team at the THW, I enjoy working with other people and I play in a football team.

JOHN Right. Do you have any other hobbies?

MAIK Yes, I do. I like building computers and robots.

JOHN Good. What are your strengths?

MAIK I'm hard-working, punctual and reliable. If I need help, I ask for it and I'm happy to help other people, too.

JOHN And your weaknesses?

MAIK I don't have much experience, because as a trainee you have to do the routine work under supervision. I need more experience, really.

JOHN I see. Why did you decide to become an electrician?

MAIK It's in the family. My dad is an electrician, so I know what it means to be one.

JOHN And why do you want this job in particular?

MAIK I want experience with a big company outside Germany, because my traineeship is only the beginning. My brother worked in Dubai for two years and he learned a lot, so I want to do the same sort of thing. I've got a lot of information on Abu Dhabi and the United Arab Emirates and now I want to go and see for myself.

JOHN Why should we employ you and not someone else?

MAIK I expect to get a good final grade at the end of my training and I really want to work somewhere very different.

JOHN All right. What are your plans for the future?

MAIK I want to set up my own business. I'll need to be a master electrician, but first I want to see the world.

JOHN I see. Do you have any questions you'd like to ask?

MAIK Yes, I do. What age are the other employees in Abu Dhabi and what countries are they from?

UNIT WORD LIST

Dieses Wörterverzeichnis enthält alle Wörter, die nicht in der Liste des Grundwortschatzes enthalten sind, d.h. die nicht als bekannt vorausgesetzt werden. t = das Wort befindet sich in den *Transcripts* (Hörverständnistexten). p = das Wort befindet sich in den *Pairwork Files*.

UNIT 1

6	**company** ['kʌmpəni]	Unternehmen, Firma
	warm up ['wɔːm ʌp]	Aufwärmphase
	wire ['waɪə]	Leitungen verlegen
	customer ['kʌstəmə]	Kunde/Kundin
	private ['praɪvət]	privat
	commercial [kə'mɜːʃl]	Geschäfts-, gewerblich
	industrial [ɪn'dʌstriəl]	Industrie-, industriell
	specialize in ['speʃəlaɪz ɪn]	sich spezialisieren auf
	employee [ɪm'plɔɪiː]	Angestellte/r
	located [ləʊ'keɪtɪd]	gelegen, mit Sitz in
	trainee [treɪ'niː]	Auszubildende/r, Praktikant/in
	qualified electrician [ˌkwɒlɪfaɪd ɪˌlek'trɪʃn]	qualifizierte/r Elektriker/in, Elektrotechniker/in
	master electrician [ˌmɑːstər ɪˌlek'trɪʃn]	Elektro(technik)-meister/in
	electronics technician [ɪlekˌtrɒnɪks te'knɪʃn]	Elektroniker/in
	contract ['kɒntrækt]	Vertrag
	businessman ['bɪznəsmən]	Geschäftsmann
	paint [peɪnt]	Farbe, Lack
	distribution [ˌdɪstrɪ'bjuːʃn]	Vertrieb, (Aus-)Lieferung
	receive [rɪ'siːv]	erhalten, bekommen
	following ['fɒləʊɪŋ]	folgende
	statement ['steɪtmənt]	Aussage
7	**electrical** [ɪ'lektrɪkl]	elektrisch, Strom-
	installation [ˌɪnstə'leɪʃn]	Installation, Einbau
	architect ['ɑːkɪtekt]	Architekt/in
	contractor [kən'træktə]	Unternehmer (mit Werkvertrag), Auftragnehmer/in
	excellent ['eksələnt]	ausgezeichnet, hervorragend
	reputation [ˌrepju'teɪʃn]	Ruf
	perform [pə'fɔːm]	aus-, durchführen
	on schedule [ɒn 'ʃedjuːl]	im Zeitplan, planmäßig
	satisfaction [ˌsætɪs'fækʃn]	Zufriedenheit
	experience [ɪk'spɪəriəns]	(Berufs-)Erfahrung
	business ['bɪznəs]	Geschäft
	be delighted [bi dɪ'laɪtɪd]	sich (sehr) freuen
	discuss [dɪ'skʌs]	besprechen, erörtern
	further ['fɜːðə]	weitere
	detail ['diːteɪl]	Einzelheit, Detail
	in person [ɪn 'pɜːsn]	persönlich
	arrange [ə'reɪndʒ]	vereinbaren, verabreden
	look forward to [ˌlʊk 'fɔːwəd tə]	sich freuen auf
	Yours sincerely [jɔːz sɪn'sɪəli]	Mit freundlichen Grüßen
	managing director [ˌmænɪdʒɪŋ də'rektə]	Geschäftsführer/in
8	**introduce** [ˌɪntrə'djuːs]	(sich) vorstellen, (miteinander) bekannt machen
	shake hands [ʃeɪk 'hændz]	sich die Hand geben
	set up [ˌset 'ʌp]	eröffnen, einrichten
	storage ['stɔːrɪdʒ]	Lager(haltung)
	space [speɪs]	Platz, Fläche
	lighting ['laɪtɪŋ]	Beleuchtung, Licht
	warehouse ['weəhaʊs]	Lager(halle, -haus)
	reduce [rɪ'djuːs]	verringern, senken
	packing ['pækɪŋ]	Verpackung, Pack-
	error ['erə]	Fehler, Irrtum, Versehen
	certainly ['sɜːtnli]	sicherlich, gewiss, bestimmt
	explosion-proof [ɪk'spləʊʒn pruːf]	explosionssicher
	fittings ['fɪtɪŋz]	Anschlüsse
	regulation [ˌregju'leɪʃn]	Verordnung, Vorschrift
	switch [swɪtʃ]	Schalter
	plug [plʌg]	Stecker
	lighting element [ˌlaɪtɪŋ 'elɪmənt]	Leuchtelement
	airtight ['eətaɪt]	luftdicht
	spark [spɑːk]	Funke
	set off [ˌset 'ɒf]	auslösen, hervorrufen
	sound [saʊnd]	anhören, klingen
	choice [tʃɔɪs]	Wahl
	case [keɪs]	Fall
	petrol station ['petrəl steɪʃn]	Tankstelle
	site [saɪt]	(Bau-)Stelle
	install [ɪn'stɔːl]	installieren, einbauen
	complete [kəm'pliːt]	vervollständigen, ergänzen
	happen ['hæpən]	passieren, geschehen
9	**phrase** [freɪz]	(Rede-)Wendung, Ausdruck
	definition [ˌdefɪ'nɪʃn]	Definition
	gap [gæp]	Lücke
	accept [ək'sept]	an-, übernehmen
	expect [ɪk'spekt]	erwarten, schätzen, annehmen
	goods [gʊdz]	Ware(n), Güter
	store [stɔː]	lagern, aufbewahren
	send out [ˌsend 'aʊt]	verschicken, versenden,

		ausliefern
	cause [kɔːz]	verursachen, auslösen
	socket ['sɒkɪt]	Steckdose
	distance ['dɪstəns]	Entfernung, Strecke
	mobile ['məʊbaɪl]	Handy, Mobilfunk
	sales division [ˌseɪlz dɪ'vɪʒn]	Verkaufsabteilung
	report [rɪ'pɔːt]	Bericht
10	**mix up** [ˌmɪks 'ʌp]	verwechseln, vermischen, durcheinander bringen
	solution [sə'luːʃn]	Lösung
	network technology [ˌnetwɜːk tek'nɒlədʒi]	Netzwerktechnik
	energy ['enədʒi]	Energie
	sun [sʌn]	Sonne
	PV (photovoltaic) system [ˌpiː 'viː ˌfəʊtə ʊvəʊlˌteɪɪk 'sɪstəm]	Photovoltaik-Anlage
	solar technology [ˌsəʊlə tek'nɒlədʒi]	Solartechnik
	hotline ['hɒtlaɪn]	Hotline, Kundentelefon
	put in touch [pʊt ɪn 'tʌtʃ]	verbinden
	solve [sɒlv]	lösen
	without delay [wɪðaʊt dɪ'leɪ]	unverzüglich
	communications [kəmjuːnɪˌkeɪʃnz tek'nɒlədʒi]	Nachrichtentechnik technology
	state-of-the-art [ˌsteɪt əv ði 'ɑːt]	auf dem neuesten Stand
	know-how ['nəʊhaʊ]	Fachkenntnis, Sachverstand
	rewire [ˌriː'waɪə]	neu verkabeln
	modernize ['mɒdənaɪz]	modernisieren
	lighting technology [ˌlaɪtɪŋ tek'nɒlədʒi]	Lichttechnik
	fibre optic [ˌfaɪbər 'ɒptɪk]	Glasfaser
	copper ['kɒpə]	Kupfer
	cable ['keɪbl]	Kabel, Leitung
	service ['sɜːvɪs]	warten, pflegen
	after-sales service ['ɑːftəseɪlz sɜːvɪs]	Kundendienst, Wartung
	reliable [rɪ'laɪəbl]	zuverlässig
	afford [ə'fɔːd]	(es) sich leisten (können)
	DVB digital TV [ˌdiː viː 'biː 'dɪdʒɪtl ˌtiː 'viː]	digitaler Fernsehempfang
	DAB digital radio [ˌdiː eɪ 'biː 'dɪdʒɪtl 'reɪdiəʊ]	digitaler Rundfunkempfang
	solar energy [ˌsəʊlə 'enədʒi]	Solarenergie
	cell [sel]	Zelle
	amount [ə'maʊnt]	Summe, Betrag
	modern ['mɒdn]	modern
	van [væn]	(Werkzeug-)Wagen
	break down [ˌbreɪk 'daʊn]	kaputt gehen
	whole [həʊl]	ganz
11	**organization chart** [ˌɔːgənaɪ'zeɪʃn tʃɑːt]	Organisationsübersicht, -struktur
	employ [ɪm'plɔɪ]	beschäftigen
	administrative [əd'mɪnɪstrətɪv]	Verwaltungs-
	technical ['teknɪkl]	technisch
	section ['sekʃn]	Abteilung
	director [də'rektə]	Direktor/in, Leiter/in
	under ['ʌndə]	unter
	warehouseman ['weəhaʊsmən]	Lagerist/in
	in charge of [ɪn 'tʃɑːdʒ əv]	verantwortlich für
	household ['haʊshəʊld]	Haushalt(s-)
	wiring ['waɪərɪŋ]	Verkabelung
	repair [rɪ'peə]	reparieren
	electric [ɪ'lektrɪk]	elektrisch, Elektro-
	motor ['məʊtə]	Motor
	purchase, purchasing ['pɜːtʃəs, -ɪŋ]	(Ein-)Kauf, Einkaufs-
	assistant [ə'sɪstənt]	Assistent/in
	human resources [ˌhjuːmən rɪ'sɔːsɪz]	Personal(abteilung)
	accounts [ə'kaʊnts]	Buchhaltung
	warehouse clerk ['weəhaʊs klɑːk]	Lagerarbeiter/in
12	**department** [dɪ'pɑːtmənt]	Abteilung
	monotonous [mə'nɒtənəs]	monoton
	hire ['haɪə]	ein-, anstellen
	fire ['faɪə]	feuern, entlassen
	stressful ['stresfl]	(sehr) anstrengend, stressig
	connect (up) [kəˌnekt 'ʌp]	(miteinander) verbinden, anschließen
	LAN (local area network) [læn, ləʊkl ˌeəriə 'netwɜːk]	lokales Netz
	cellar ['selə]	Keller
	immediately [ɪ'miːdiətli]	sofort, umgehend
	D.I.Y. (Do-it-Yourself) [ˌdiː aɪ 'waɪ, ˌdu ɪt jɔː'self]	Heimwerker(-)
	cordless phone ['kɔːdləs fəʊn]	schnurloses Telefon
	on call [ɒn 'kɔːl]	auf Abruf, abrufbereit
	bleeper ['bliːpə]	(Telefon-)Pieper
	invent [ɪn'vent]	erfinden
	description [dɪ'skrɪpʃn]	Beschreibung, Schilderung
	guess [ges]	(er)raten
	draw up [ˌdrɔː 'ʌp]	abfassen, aufstellen
	point [pɔɪnt]	Punkt
	presentation [ˌprezn'teɪʃn]	Vorführung
	based in ['beɪst ɪn]	mit (Haupt-)Sitz in
	city ['sɪti]	(Groß-, Innen-)Stadt
	local ['ləʊkl]	örtlich, lokal
	including [ɪn'kluːdɪŋ]	einschließlich, inbegriffen, darunter

	engineer [ˌendʒɪ'nɪə]	Ingenieur/in, Techniker/in
	staff [stɑ:f]	Angestellte, Personal
	factory ['fæktəri]	Fabrik
	generator ['dʒenəreɪtə]	Generator
	equipment [ɪ'kwɪpmənt]	Ausrüstung, Geräte, Einrichtung
13	**link** [lɪŋk]	Verbindung, Link
	guided tour [ˌgaɪdɪd 'tʊə]	Besichtigung, Führung
	diagram ['daɪəgræm]	grafische Darstellung, Grafik
	maintenance ['meɪntənəns]	Wartung
	delivery [dɪ'lɪvəri]	(Aus-)Lieferung, Liefer-
	vehicle ['vi:əkl]	Fahrzeug
	administration [ədˌmɪnɪ'streɪʃn]	Verwaltung
	reception [rɪ'sepʃn]	Empfang
	communication [kəˌmju:nɪ'keɪʃn]	Kommunikation(s-)
	network ['netwɜ:k]	Netz(werk)
14	**popular** ['pɒpjələ]	beliebt, populär
	nowadays ['naʊədeɪz]	heutzutage
	unqualified [ˌʌn'kwɒlɪfaɪd]	nicht qualifiziert
	aerial ['eəriəl]	Antenne
	stock [stɒk]	Vorrat, Lager(bestand)
	supply [sə'plaɪ]	(Waren-)Angebot
	show around [ˌʃəʊ ə'raʊnd]	herumführen
	English-speaking ['ɪŋglɪʃspi:kɪŋ]	englischsprachig
	act out [ˌækt 'aʊt]	vorspielen, aufführen
	expression [ɪk'spreʃn]	Ausdruck
	corridor ['kɒrɪdɔ:]	Gang, Flur, Korridor
	give directions [gɪv də'rekʃnz]	den Weg beschreiben
	on your left [ɒn jɔ: 'left]	zur Linken
	next door to [ˌnekst 'dɔ: tə]	neben
	in front of [ɪn 'frʌnt əv]	vor
	at the end of [ət ði 'end əv]	am Ende des/der
	on the other side of [ɒn ði 'ʌðə saɪd əv]	gegenüber
15	**apply** [ə'plaɪ]	sich bewerben
	online ['ɒnlaɪn]	online, im/übers Internet
	location [ləʊ'keɪʃn]	(Stand-)Ort
	rate [reɪt]	Rate, (Stunden-)Satz
	require [rɪ'kwaɪə]	suchen
	maintenance electrician [ˌmeɪntənəns ɪˌlek'trɪʃn]	Wartungstechniker/in
	candidate ['kændɪdət]	Bewerber/in
	fabricate ['fæbrɪkeɪt]	herstellen, fertigen
	panel ['pænl]	Anzeige-, Schalttafel
	include [ɪn'klu:d]	umfassen, einschließen
	circuit ['sɜ:kɪt]	Schaltung, (Strom-)Kreis
	work with ['wɜ:k wɪð]	arbeiten mit
	voltage ['vəʊltɪdʒ]	Spannung
	preferably ['prefərəbli]	am besten, vorzugsweise
	shift [ʃɪft]	Schicht
	consider [kən'sɪdə]	betrachten (als)
	overtime ['əʊvətaɪm]	Überstunden
	submit [səb'mɪt]	einreichen
	résumé ['rezju:meɪ]	(tabellarischer) Lebenslauf
	for consideration [fə kənˌsɪdə'reɪʃn]	zur Beurteilung
	invite [ɪn'vaɪt]	einladen
	interview ['ɪntəvju:]	Vorstellungsgespräch
	required [rɪ'kwaɪəd]	erforderlich
	skill [skɪl]	Fertigkeit, Fähigkeit
	electronic [ˌɪlek'trɒnɪk]	elektronisch
	term [tɜ:m]	Bedingung
	employment [ɪm'plɔɪmənt]	Arbeit, Beschäftigung
	comprehensive [ˌkɒmprɪ'hensɪv]	umfassend
	medical ['medɪkl]	medizinisch
	dental ['dentl]	zahnärztlich
	optical ['ɒptɪkl]	optisch, augenärztlich
	benefit ['benɪfɪt]	Unterstützung
	optional ['ɒpʃənl]	wahlweise
	government-sponsored ['gʌvənmənt spɒnsəd]	staatlich gefördert
	pension ['penʃn]	Rente, Pension
	competitive [kəm'petətɪv]	konkurrenzfähig
	wage [weɪdʒ]	Lohn, Gehalt
	solid ['sɒlɪd]	fest, solide
	career [kə'rɪə]	beruflich(er Werdegang), Karriere
	advancement [əd'vɑ:nsmənt]	Aufstieg
	opportunity [ˌɒpə'tju:nəti]	Chance, Möglichkeit
	click [klɪk]	(an)klicken
	avionics [ˌeɪvi'ɒnɪks]	Avionik, Flugzeugelektronik
	engineering [ˌendʒɪ'nɪərɪŋ]	Technik, Maschinenbau
	compare [kəm'peə]	vergleichen, sich vergleichen lassen
	training course ['treɪnɪŋ kɔ:s]	Lehrgang
	cover ['kʌvə]	(ab)decken, einschließen
	discussion [dɪ'skʌʃn]	Diskussion, Gespräch, Erörterung
	disadvantage [ˌdɪsəd'vɑ:ntɪdʒ]	Nachteil
	be interested in [bi 'ɪntrəstɪd ɪn]	interessiert sein an, sich interessieren für
	ideal [aɪ'di:əl]	ideal

UNIT 2

16	**workplace** ['wɜ:kpleɪs]	Arbeitsplatz
	installation electrician [ˌɪnstəˌleɪʃn ɪˌlek'trɪʃn]	Elektroinstallateur/in

	panel building electrician [ˌpænl ˌbɪldɪŋ ɪˌlekˈtrɪʃn]	Schalttafel(bau)-elektriker/in
	repair and maintenance electrician [rɪˌpeər ən ˌmeɪntənəns ɪˌlekˈtrɪʃn]	Elektrotechniker/in für Reparatur und Wartung
	street lighting electrician [ˌstriːt laɪt ɪˌlekˈtrɪʃn]	Elektriker/in für öffentliche Beleuchtung
	involve [ɪnˈvɒlv]	einschließen, umfassen
	assemble [əˈsembl]	zusammensetzen, -bauen, montieren
	domestic household [dəˌmestɪk ˈhaʊshəʊld]	Privathaushalt
	inspect [ɪnˈspekt]	kontrollieren, überprüfen
	industrial plant [ɪnˈdʌstriəl plɑːnt]	Industriewerk, -anlage
	maintain [meɪnˈteɪn]	warten, pflegen
	decide [dɪˈsaɪd]	(sich) entscheiden
17	**wiring plan** [ˈwaɪərɪŋ plæn]	Verdrahtungs-, Leitungsplan
	control panel [kənˈtrəʊl pænl]	Kontrollpult, -tafel
	PLC (programmable logic controller) [ˌpiː el ˈsiː, ˌprəʊgræməbl ˌlɒdʒɪk kənˈtrəʊlə]	speicherprogrammierbare Steuerung
	control [kənˈtrəʊl]	steuern
	heating [ˈhiːtɪŋ]	Heizung
	air conditioning [ˈeə kəndɪʃnɪŋ]	Klimaanlage
	ventilation system [ˌventɪˈleɪʃn sɪstəm]	Belüftungssystem
	workshop [ˈwɜːkʃɒp]	Werkstatt, Werkshalle
	college [ˈkɒlɪdʒ]	(Berufs-)Fachschule, (Fach-)Hochschule
	complex [ˈkɒmpleks]	kompliziert, komplex
	strip out [ˌstrɪp ˈaʊt]	demontieren, herausnehmen, entfernen
	replace with [rɪˈpleɪs wɪð]	ersetzen durch, austauschen gegen
	drawing [ˈdrɔːɪŋ]	Zeichnung
	bend [bend]	biegen
	conduit [ˈkɒndjuɪt]	Kabelrohr
	fix [fɪks]	befestigen
	enjoy [ɪnˈdʒɔɪ]	gern haben/tun
	variety [vəˈraɪəti]	Vielfalt, Abwechslung
	block of flats [blɒk əv ˈflæts]	(Wohn-)Block, Hochhaus
	travel around [ˌtrævl əˈraʊnd]	herumfahren
	depending on [dɪˈpendɪŋ ɒn]	je nachdem
	working week [ˌwɜːkɪŋ ˈwiːk]	Arbeitswoche
	responsible for [rɪˈspɒnsəbl fə]	verantwortlich für, zuständig für
	occur [əˈkɜː]	vorkommen, eintreten
	item [ˈaɪtəm]	Gegenstand, Ding
	circuit breaker [ˈsɜːkɪt breɪkə]	Stromunterbrecher, Überspannungsschalter
	fuse [fjuːz]	Sicherung
	periodical(ly) [ˌpɪəriˈɒdɪkl, -kli]	regelmäßig
	operate [ˈɒpəreɪt]	arbeiten, laufen, funktionieren
	properly [ˈprɒpəli]	richtig, korrekt
	locate [ləʊˈkeɪt]	ausfindig machen
	task [tɑːsk]	Aufgabe, Arbeit
	troubleshoot [ˈtrʌblʃuːt]	Störung suchen
	function [ˈfʌŋkʃn]	funktionieren, arbeiten
	available [əˈveɪləbl]	anwesend, verfügbar
	deal with [ˈdiːl wɪð]	sich befassen/beschäftigen mit, bearbeiten, erledigen
	diagnose [ˈdaɪəgnəʊz]	diagnostizieren
	traffic [ˈtræfɪk]	(Straßen-)Verkehr
	traffic light [ˈtræfɪk laɪt]	Ampel
	motorway [ˈməʊtəweɪ]	Autobahn
	message [ˈmesɪdʒ]	Mitteilung, Hinweis
	sign [saɪn]	(Verkehrs-, Hinweis-)Schild
	high access platform [haɪ ˌækses ˈplætfɔːm]	Arbeits-, Hebebühne
	mostly [ˈməʊstli]	hauptsächlich, meistens
	in the open air [ɪn ði ˌəʊpən ˈeə]	im Freien, an der frischen Luft
	height [haɪt]	(große) Höhe
	mind [maɪnd]	etw dagegen haben
	in fact [ɪn ˈfækt]	tatsächlich, eigentlich
	working conditions [ˌwɜːkɪŋ kənˈdɪʃnz]	Arbeitsbedingungen
	inconvenient [ˌɪnkənˈviːniənt]	unbequem
	especially [ɪˈspeʃəli]	besonders
	condition [kənˈdɪʃn]	Bedingung
	fixed [fɪkst]	fest(gesetzt)
	emergency [ɪˈmɜːdʒənsi]	Notfall
	work site [ˈwɜːk saɪt]	Baustelle
	a set of [ə ˈset əv]	eine Reihe von
	malfunction [ˌmælˈfʌŋkʃn]	nicht funktionieren
18	**carry out** [ˌkæri ˈaʊt]	aus-, durchführen
	routine [ruːˈtiːn]	Routine(arbeit), routinemäßig
	strange [streɪndʒ]	ungewöhnlich
	equivalent [ɪˈkwɪvələnt]	Entsprechung
	bracket [ˈbrækɪt]	Klammer
	production line [prəˈdʌkʃn laɪn]	Fließ-, Fertigungsband
	power station [ˈpaʊə steɪʃn]	Kraftwerk
	oil refinery [ˌɔɪl rɪˈfaɪnəri]	Ölraffinerie
	indoors [ˈɪndɔːz]	drinnen, in geschlossenen Räumen

	comparison [kəm'pærɪsn]	Vergleich
	pick [pɪk]	auswählen
	job profile ['dʒɒb prəʊfaɪl]	Stellenbeschreibung, Berufsbild
	similar ['sɪmələ]	ähnlich
	device [dɪ'vaɪs]	Gerät, Apparat
	component [kəm'pəʊnənt]	(Bau-)Element, (Bau-)Teil
	advise [əd'vaɪz]	beraten
	technical drawing [ˌteknɪkl 'drɔːɪŋ]	technische Zeichnung
	instruction [ɪn'strʌkʃn]	Anweisung, (Betriebs-)Anleitung
19	**equal** ['iːkwəl]	gleich (groß)
	in favour of [ɪn 'feɪvər əv]	für (etwas)
	protect [prə'tekt]	(be)schützen
	prepare (to) [prɪ'peə tə]	(sich) vorbereiten (auf)
	brainstorm ['breɪnstɔːm]	(Ideen) sammeln
	argument ['ɑːgjumənt]	Argument
	support [sə'pɔːt]	unterstützen
	point of view [ˌpɔɪnt əv 'vjuː]	Standpunkt, Ansicht
	web page ['web peɪdʒ]	Internetseite
	ever ['evə]	je(mals), schon (ein)mal
	point out [ˌpɔɪnt 'aʊt]	hinweisen auf, aufmerksam machen auf
	agree with [ə'griː wɪð]	zustimmen
	absolutely ['æbsəluːtli]	absolut, völlig
	doubt [daʊt]	zweifeln
	disagree [ˌdɪsə'griː]	anderer Meinung sein, widersprechen, nicht zustimmen
	be wrong about [bi 'rɒŋ əbaʊt]	Unrecht haben mit/bei
	agree to disagree [əˌgriː tə ˌdɪsə'griː]	sich darauf einigen, dass man sich nicht einigen kann
20	**workbench** ['wɜːkbentʃ]	Werkbank
	tool [tuːl]	Werkzeug, Instrument, Gerät
	Allen keys ['ælən kiːz]	Inbusschlüssel, Innensechskant-schlüssel
	water pump pliers [ˌwɔtə pʌmp 'plaɪəz]	Wasserpumpenzange
	power drill ['paʊə drɪl]	elektrische Bohrmaschine, Elektrobohrer
	folding rule ['fəʊldɪŋ ruːl]	Zollstock
	hacksaw ['hæksɔː]	Bügelsäge
	hammer ['hæmə]	Hammer
	multimeter [ˌmʌlti'miːtə]	Vielfachmessinstrument
	Phillips screwdriver [ˌfɪlɪps 'skruːdraɪvə]	Kreuzschlitz-schraubendreher
	round nose pliers [ˌraʊnd nəʊz 'plaɪəz]	Rundzange
	flat-bladed screwdriver ['flætbleɪdɪd 'skruːdraɪvə]	Schlitzschrauben-zieher
	side-cutting pliers ['saɪdkʌtɪŋ plaɪəz]	Seitenschneider
	soldering iron ['səʊldərɪŋ aɪən]	Lötkolben
	spanner ['spænə]	Schraubenschlüssel
	needle nose pliers [ˌniːdl nəʊz 'plaɪəz]	Spitzzange
	vernier calliper ['vɜːniə kælɪpə]	Messschieber
	wire strippers ['waɪə strɪpəz]	Abisolierzange
21	**tighten** ['taɪtn]	fest-, anziehen, festschrauben
	screw [skruː]	Schraube
	drill [drɪl]	bohren
	bolt [bəʊlt]	Riegel, Bolzen, Schraube
	grip [grɪp]	(er)greifen
	measure ['meʒə]	(ver)messen
	electrical circuit [ɪˌlektrɪkl 'sɜːkɪt]	Stromkreis
	hole [həʊl]	Loch
	a pair of [ə 'peər əv]	ein Paar, eine (Zange)
	solder ['səʊldə]	löten
	nut [nʌt]	(Schrauben-)Mutter
	strip [strɪp]	abisolieren
	order ['ɔːdə]	bestellen
21t	**place an order** [ˌpleɪs ən 'ɔːdə]	einen Auftrag erteilen, eine Bestellung aufgeben, bestellen
	sir [sɜː]	mein Herr
	in stock [ɪn 'stɒk]	vorrätig, auf Lager
	invoice ['ɪnvɔɪs]	eine Rechnung ausstellen (auf), berechnen
	credit card payment [ˌkredɪt kɑːd 'peɪmənt]	Bezahlung mit Kreditkarte
	cash on delivery [ˌkæʃ ɒn dɪ'lɪvəri]	Barzahlung bei Lieferung, Barkauf
21	**pairwork** ['peəwɜːk]	Partnerarbeit
	file [faɪl]	Akte, (Akten-)Ordner
21p	**DMM (digital multimeter)** [ˌdiː em 'em, ˌdɪdʒɪtl ˌmʌlti'miːtə]	digitales Vielfachmessinstrument
	salesperson ['seɪlzpɜːsn]	Verkäufer/in
	cordless drill ['kɔːdləs drɪl]	Akku-Schrauber, -Bohrer
	payment terms ['peɪmənt tɜːmz]	Zahlungsbedingungen
	electric drill [ɪ'lektrɪk drɪl]	Elektrobohrer
22	**training** ['treɪnɪŋ]	Ausbildung
	leave [liːv]	abgehen (von), verlassen
	apprenticeship [ə'prentɪʃɪp]	Lehre, Ausbildung

	practical ['præktɪkl]	praktisch
	day release [ˌdeɪ rɪ'liːs]	freier Tag (zur Fortbildung)
	pass [pɑːs]	(Examen) bestehen, ablegen
	NVQ (National Vocational Qualification) [ˌen viː 'kjuː, ˌnæʃnəl vəʊˌkeɪʃənl ˌkwɒlɪfɪ'keɪʃn]	staatlicher Berufsabschluss in GB
	straight away [streɪt ə'weɪ]	sofort, gleich, auf der Stelle
	grinding machine ['graɪndɪŋ məʃiːn]	Schleifmaschine
	develop [dɪ'veləp]	entwickeln, bekommen
	broken ['brəʊkən]	kaputt, defekt
23	**roller bearing** ['rəʊlə beərɪŋ]	Kugellager
	restart [ˌriː'stɑːt]	neu starten
	as if [əz 'ɪf]	als ob, als wenn
	contact ['kɒntækt]	Kontakt
	contactor ['kɒntæktə]	Kontakt-, Impulsgeber
	be designed to [bi dɪ'zaɪnd tə]	zu etw dienen
	break [breɪk]	(unter)brechen
	current ['kʌrənt]	Strom
	load [ləʊd]	Last, Belastung
	serve as ['sɜːv əz]	dienen als
	security gate [sɪ'kjʊərəti geɪt]	Sicherheitstor, -ventil
	faulty ['fɔːlti]	fehlerhaft, defekt
	spare [speə]	Ersatz-
	press [pres]	drücken
	start button ['stɑːt bʌtn]	Startknopf
	inverter [ɪn'vɜːtə]	Wechselrichter
	rotating speed [rəʊ'teɪtɪŋ spiːd]	Drehgeschwindigkeit
	power output [ˌpaʊər 'aʊtpʊt]	Strommenge, -stärke
	reading ['riːdɪŋ]	Anzeige
	function test ['fʌŋkʃn test]	Funktionsprüfung
	smooth(ly) ['smuːð, -li]	ruhig, rund
	armature winding resistance [ˌɑːmətʃə ˌwaɪndɪŋ rɪ'zɪstəns]	Ankerwicklungswiderstand
	overloaded [ˌəʊvə'ləʊdɪd]	überlastet
	be worth [bi 'wɜːθ]	wert sein, lohnen
24	**make notes** [meɪk 'nəʊts]	(sich) Notizen machen
	zigzag ['zɪgzæg]	zickzackförmig
	electrician [ɪˌlek'trɪʃn]	Elektriker/in, Elektrotechniker/in
	operator ['ɒpəreɪtə]	Maschinist/in
	take a look at [teɪk ə 'lʊk ət]	(sich) ansehen
25	**take part in** [teɪk 'pɑːt ɪn]	teilnehmen an, mitmachen bei
	scheme [skiːm]	Programm, Projekt
	combine [kəm'baɪn]	verbinden, kombinieren
	helpline ['helplaɪn]	Telefonberatung, telefonische Hilfe
	suitable (-ly) ['suːtəbl, -bli]	geeignet, passend
	employer [ɪm'plɔɪə]	Arbeitgeber/in, Unternehmer/in
	alternatively [ɔːl'tɜːnətɪvli]	als Alternative, als weitere Möglichkeit
	get in touch with [get ɪn 'tʌtʃ wɪð]	(sich) in Verbindung setzen mit
	directly [də'rektli]	direkt
	power generation [ˌpaʊə ˌdʒenə'reɪʃn]	Stromerzeugung
	electronic servicing [ˌɪlekˌtrɒnɪk 'sɜːvɪsɪŋ]	elektronische Wartung
	apprentice [ə'prentɪs]	Lehrling, Auszubildende/r
	time off (work) [ˌtaɪm 'ɒf wɜːk]	(arbeits)freie Zeit, arbeitsfrei
	exact [ɪg'zækt]	genau
	balance ['bæləns]	Verhältnis
	depend [dɪ'pend]	abhängen (von)
	several ['sevrəl]	einige, mehrere
	block release [ˌblɒk rɪ'liːs]	Blockunterricht
	basic ['beɪsɪk]	Grund-
	training allowance ['treɪnɪŋ əlaʊəns]	Ausbildungsbeihilfe
	guarantee [ˌgærən'tiː]	garantieren
	participant [pɑː'tɪsɪpənt]	Teilnehmer/in
	qualification [ˌkwɒlɪfɪ'keɪʃn]	Qualifikation, (Abschluss einer) Ausbildung
	package ['pækɪdʒ]	Bündel
	BTEC (National Diploma or other technical certificate) ['biː tek, ˌnæʃnəl dɪ'pləʊmə ɔːr ʌðə ˌteknɪkl sə'tɪfɪkət]	staatlicher technischer Abschluss in GB
	Key Skills qualification [ˌkiː skɪlz ˌkwɒlɪfɪ'keɪʃn]	Bescheinigung über Schlüsselqualifikationen
	problem solving ['prɒbləm sɒlvɪŋ]	Problemlösung
	commitment [kə'mɪtmənt]	Einsatz, Engagement
	reward [rɪ'wɔːd]	Belohnung, Lohn
	demand (for) [dɪ'mɑːnd fə]	Nachfrage (für), Bedarf (an)
	inform [ɪn'fɔːm]	informieren, unterrichten
	classmate ['klɑːsmeɪt]	Klassenkamerad/in
	outline ['aʊtlaɪn]	Überblick, Abriss

UNIT 3

26	**vocabulary** [və'kæbjələri]	Vokabular, Wortschatz
	consumer unit [kən'sju:mə ju:nɪt]	Sicherungskasten
	earth terminal ['ɜ:θ tɜ:mɪnl]	Erdpol, Masseklemme
	electricity meter [ɪˌlek'trɪsəti mi:tə]	Arbeitszähler
	flex [fleks]	Netzkabel
	junction box ['dʒʌŋkʃn bɒks]	Verteilerkasten
	light switch ['laɪt swɪtʃ]	Lichtschalter
	RCD (Residual Current Device) [ˌɑ: si: 'di:, rɪˌzɪdjuəl ˌkʌrənt dɪ'vaɪs]	Fehlerstromschutz-schalter
	service entrance ['sɜ:vɪs entrəns]	Wartungs-, Inspektionszugang, -anschluss
	designer [dɪ'zaɪnə]	Konstrukteur/in, Designer/in
	electrical contractor [ɪˌlektrɪkl kən'træktə]	Elektroinstallateur
	doorbell ['dɔ:bel]	Türklingel
27	**disaster** [dɪ'zɑ:stə]	Katastrophe
	cooker ['kʊkə]	Herd
	microwave ['maɪkrəweɪv]	Mikrowelle
	nowhere ['nəʊweə]	kein Platz/Ort
	attach [ə'tætʃ]	anschließen
	gas cooker ['gæs kʊkə]	Gasherd
	satellite TV ['sætəlaɪt ti:vi:]	Satellitenfernsehen
	unscrew [ˌʌn'skru:]	heraus-, abschrauben
	channel ['tʃænl]	Kanal, Leerrohr, Elektroinstallationsrohr
	style [staɪl]	Stil
	in those days [ɪn ðəʊz 'deɪz]	damals, in jener Zeit
	be not allowed [bi nɒt ə'laʊd]	nicht dürfen
	missing ['mɪsɪŋ]	fehlend, nicht vorhanden
	earth wire ['ɜ:θ waɪə]	Schutzleiter
	protective [prə'tektɪv]	Schutz-
	earth conductor [ˌɜ:θ kən'dʌktə]	Schutzleiter
	appliance [ə'plaɪəns]	Gerät
	earth [ɜ:θ]	Erde; erden
	awful ['ɔ:fl]	furchtbar, schrecklich
	pipe [paɪp]	Rohr(leitung)
	shock [ʃɒk]	Schock, Schlag
	suggest [sə'dʒest]	vorschlagen
	completely [kəm'pli:tli]	vollständig
	electric cooker [ɪˌlektrɪk 'kʊkə]	Elektroherd
	fuse box ['fju:z bɒks]	Sicherungskasten
	mention ['menʃn]	erwähnen
	colour ['kʌlə]	Farbe
28	**translation** [træns'leɪʃn]	Übersetzung
29	**terminal** ['tɜ:mɪnl]	Ausgang, Anschluss
	switch off [ˌswɪtʃ 'ɒf]	ab-, ausschalten
	lock [lɒk]	schließen
	tag [tæg]	kennzeichnen, etikettieren
	cut away [ˌkʌt ə'weɪ]	abschneiden
	insulation [ˌɪnsju'leɪʃn]	Isolierung, Isolation
	length [leŋθ]	Länge
	mounting ['maʊntɪŋ]	Montage
	recess [rɪ'ses]	Vertiefung
	plastic ['plæstɪk]	Plastik, Kunststoff
	select [sɪ'lekt]	auswählen
	two-way switch [ˌtu: weɪ 'swɪtʃ]	Wechselschalter
	part of ['pɑ:t əv]	Teil von
30	**protective earth** [prə'tektɪv ɜ:θ]	Schutzleiter
	power (a machine) [ˌpaʊər ə mə'ʃi:n]	(ein Gerät) antreiben
	known as ['nəʊn əz]	bekannt als
	escape [ɪ'skeɪp]	entweichen
	separate ['seprət]	getrennt
	pass through ['pɑ:s θru:]	durchlaufen, passieren
	talk to ['tɔ:k tə]	sprechen mit
	choose [tʃu:z]	(aus)wählen
30t	**fittings** ['fɪtɪŋz]	Anschlüsse
	ceiling light ['si:lɪŋ laɪt]	Deckenleuchte
	single socket [ˌsɪŋgl 'sɒkɪt]	einfache Steckdose
	mark [mɑ:k]	markieren, kennzeichnen
	else [els]	sonst (noch)
	corner ['kɔ:nə]	Ecke
	five-core cable [ˌfaɪv kɔ: 'keɪbl]	fünfpolige Leitung
	extractor fan [ɪk'stræktə fæn]	Abzugshaube
	fridge-freezer combination [ˌfrɪdʒ 'fri:zə ˌkɒmbɪ'neɪʃn]	Gefrierkombination
	dishwasher ['dɪʃwɒʃə]	Geschirrspüler
	sink unit ['sɪŋk ju:nɪt]	Spüle, Spülbecken
	triple ['trɪpl]	dreifach, Dreifach-
	double ['dʌbl]	doppelt, Doppel-
	single ['sɪŋgl]	einfach
	rush [rʌʃ]	sich beeilen
30	**spot** [spɒt]	herausfinden, bemerken
	refer to [rɪ'fɜ: tə]	hier: vergleichen mit
	necessary ['nesəsəri]	nötig, erforderlich
31	**outlet** ['aʊtlet]	Anschluss
	earthed socket outlet [ɜ:θt 'sɒkɪt aʊtlet]	geerdete Steckdose, Schukosteckdose
	switched earthed socket outlet [ˌswɪtʃt ɜ:θt 'sɒkɪt aʊtlet]	schaltbare geerdete Steckdose
	fourfold ['fɔ:fəʊld]	vierfach
	shaver socket ['ʃeɪvə sɒkɪt]	Steckdose für Rasierer

	switched lamp ['swɪtʃt læmp]	Leuchte mit Schalter
	spot lamp ['spɒt læmp]	Spot
	bell [bel]	Klingel, Glocke
	double pole ['dʌbl pəʊl]	zweipolig, Zweipol-
	push button ['pʊʃ bʌtn]	Drucktaste
	intermediate switch [ˌɪntə'mi:diət swɪtʃ]	Zwischenschalter
	lighting outlet [ˌlaɪtɪŋ 'aʊtlet]	Lichtanschluss
	armchair ['ɑ:mtʃeə]	Sessel
	in the middle [ɪn ðə 'mɪdl]	in der Mitte
	dressing table ['dresɪŋ teɪbl]	Frisierkommode
	next to ['nekst tə]	neben
	walk-in wardrobe [ˌwɔ:k ɪn 'wɔ:drəʊb]	begehbarer Kleiderschrank
32	**introduce oneself** [ˌɪntrə'dju:s wʌnself]	(sich) vorstellen
	role-play ['rəʊl pleɪ]	als Rollenspiel durchführen
	elderly ['eldəli]	älter
	informal [ɪn'fɔ:ml]	formlos, locker
	neutral ['nju:trəl]	neutral
	polite [pə'laɪt]	höflich
	madam ['mædəm]	gnädige Frau
	version ['vɜ:ʃn]	Version, Fassung
	appropriate [ə'prəʊpriət]	angemessen, passend
33	**final** ['faɪnl]	letzte/r/s
	businesswoman ['bɪznəswʊmən]	Geschäftsfrau
34	**ill** [ɪl]	krank
	warning ['wɔ:nɪŋ]	Warnung, Warnhinweis
	workman ['wɜ:kmən]	Handwerker
	overhead ['əʊvəhed]	an der Decke
	power cable ['paʊə keɪbl]	Stromkabel, -leitung
	ladder ['lædə]	Leiter
	get a shock [get ə 'ʃɒk]	einen Schlag bekommen
	kill oneself ['kɪl wʌnself]	sich töten
	nasty ['nɑ:sti]	schlimm
	accident ['æksɪdənt]	Unfall
	tie one's hair back [ˌtaɪ wʌnz heə 'bæk]	das Haar zusammenbinden
	trip [trɪp]	betätigen, auslösen
	save [seɪv]	retten
	fall into ['fɔ:l ɪntə]	hineinfallen
	electrocute oneself [ɪ'lektrəkju:t wʌnself]	sich durch Stromschlag töten
	turn on [ˌtɜ:n 'ɒn]	anschalten
	give an electric shock [gɪv ən ɪˌlektrɪk 'ʃɒk]	einen Stromschlag verpassen
	fall onto ['fɔ:l ɒntə]	fallen auf
	pull [pʊl]	ziehen (an)
35	**consumer** [kən'sju:mə]	Verbraucher/in
	via ['vaɪə]	über, durch
	underground ['ʌndəgraʊnd]	unterirdisch
	transport [træn'spɔ:t]	transportieren
	overland ['əʊvəlænd]	über Land
	held up by [ˌheld 'ʌp baɪ]	gehalten durch
	electricity pylon [ɪˌlek'trɪsəti paɪlən]	Strommast
	step down [ˌstep 'daʊn]	reduzieren, heruntertransformieren
	feed into ['fi:d ɪntə]	eingeben, einspeisen
	national grid [ˌnæʃnəl 'grɪd]	Energieversorgungsnetz
	step up [ˌstep 'ʌp]	hochtransformieren
	substation ['sʌbsteɪʃn]	Umspannwerk
	transformer [træns'fɔ:mə]	Transformator
	generate ['dʒenəreɪt]	erzeugen
	bullet point ['bʊlɪt pɔɪnt]	Stichpunkt

UNIT 4

36	**permanent** ['pɜ:mənənt]	ständig
	general ['dʒenrəl]	allgemein
	rule [ru:l]	Vorschrift, Regel
	water heater ['wɔ:tə hi:tə]	Wasserkocher
	recommend [ˌrekə'mend]	empfehlen
37	**ugly** ['ʌgli]	hässlich
	extension [ɪk'stenʃn]	Verlängerung
	advice [əd'vaɪs]	Rat(schlag)
	kitchenette [ˌkɪtʃɪ'net]	kleine Küche, Kochnische
	immersion heater [ɪ'mɜ:ʃn hi:tə]	Warmwasserspeicher
	tank [tæŋk]	Wasserspeicher
	continuous flow water heater [kənˌtɪnjuəs fləʊ 'wɔ:tə hi:tə]	Durchlauferhitzer
	tankless ['tæŋkləs]	ohne Speicher
	efficient [ɪ'fɪʃnt]	(leistungs)fähig
	compact ['kɒmpækt]	kompakt
	to be on the safe side [tə bi ɒn ðə 'seɪf saɪd]	um sicherzugehen
	sketch [sketʃ]	Skizze
	estimate ['estɪmət]	Kostenvoranschlag
38	**hedge** [hedʒ]	Hecke
	mow the lawn [məʊ ðə 'lɔ:n]	den Rasen mähen
	fixture ['fɪkstʃə]	(Elektro-)Installation
	converted [kən'vɜ:tɪd]	umgebaut
	undersink [ˌʌndə'sɪŋk]	Unterbau-
	radio shack [ʃæk]	Funkerbude
39	**chain reaction** [ˌtʃeɪn ri'ækʃn]	Kettenreaktion
	melt [melt]	schmelzen
	annoyed [ə'nɔɪd]	verärgert, irritiert

	get into trouble [ˌget ɪntə 'trʌbl]	Probleme bekommen
	paragraph ['pærəgrɑ:f]	Absatz, Abschnitt
40	**flow along** ['fləʊ əlɒŋ]	durchfließen, -strömen
	phase [feɪz]	Phase
	source [sɔ:s]	Quelle
	act as ['ækt əz]	fungieren als
	resistance [rɪ'zɪstəns]	Widerstand
	uncontrolled [ˌʌnkən'trəʊld]	unkontrolliert
	surge [sɜ:dʒ]	(plötzlicher) Anstieg
	massive ['mæsɪv]	riesig, enorm
	flow through ['fləʊ θru:]	durchströmen, -fließen
	heat [hi:t]	Hitze, Wärme
	basically ['beɪsɪkli]	grundsätzlich
	react (to) [ri'ækt tə]	reagieren (auf)
	whereas [ˌweər'æz]	wohingegen, während
	leakage ['li:kɪdʒ]	Austreten, Ausströmen
	ground [graʊnd]	Erdung
	flat [flæt]	Wohnung
	amp(ere) [æmp, 'æmpeə]	Ampere
	rise [raɪz]	(an)steigen
	nominal value [ˌnɒmɪnl 'vælju:]	Nennwert, Nominalwert
	figure ['fɪgə]	Abbildung
	electric shock [ɪˌlektrɪk 'ʃɒk]	elektrischer Schlag
	sudden ['sʌdn]	plötzlich
	increase ['ɪŋkri:s]	Zunahme, Steigerung
	liquid ['lɪkwɪd]	Flüssigkeit
	harm [hɑ:m]	Schaden
41	**nominal current** [ˌnɒmɪnl 'kʌrənt]	Nennstrom
	bimetallic strip [ˌbaɪmətælɪk 'strɪp]	Bimetallstreifen
	solenoid ['sɒlənɔɪd]	Solenoid, Magnetspule
	unacceptable [ˌʌnək'septəbl]	unannehmbar, inakzeptabel
	thermal-magnetic circuit-breaker [ˌθɜ:ml mægˌnetɪk 'sɜ:kɪt breɪkə]	thermisch-magnetischer Schutzschalter
	thermal circuit-breaker [ˌθɜ:ml 'sɜ:kɪt breɪkə]	thermischer Schutzschalter
	be made of [bi 'meɪd əv]	bestehen aus
	expand [ɪk'spænd]	sich ausdehnen
	temperature ['temprətʃə]	Temperatur
	straight [streɪt]	gerade
	however [haʊ'evə]	jedoch
	bottom ['bɒtəm]	Unterseite
	combination [ˌkɒmbɪ'neɪʃn]	Kombination
	advantage [əd'vɑ:ntɪdʒ]	Vorteil
	millisecond ['mɪlisekənd]	Millisekunde
	overload ['əʊvələʊd]	Überlastung
42	**coil** [kɔɪl]	Windung, Spule
	identical [aɪ'dentɪkl]	identisch
	wind [waɪnd]	(sich) winden, biegen
	cancel out [ˌkænsl 'aʊt]	sich (gegenseitig) aufheben
	lost [lɒst]	verloren (gegangen)
	neutral coil ['nju:trəl kɔɪl]	Spule im Neutralleiter
	phase coil ['feɪz kɔɪl]	Spule im Phasenleiter
	disconnect [ˌdɪskə'nekt]	unterbrechen, trennen
	residual [rɪ'zɪdʒuəl]	Rest-, restlich
	physical ['fɪzɪkl]	physikalisch
	test button ['test bʌtn]	Prüf-, Testknopf
	sensing coil ['sensɪŋ kɔɪl]	Sensorspule
	trigger ['trɪgə]	auslösen
43	**switch on** [ˌswɪtʃ 'ɒn]	einschalten
	live [laɪv]	unter Spannung/Strom (stehend)
	the mains [ðə 'meɪnz]	Stromnetz
	bother ['bɒðə]	stören, auf die Nerven gehen
	neighbour ['neɪbə]	Nachbar/in
	get the wrong end of the stick [get ðə ˌrɒŋ 'end əv ðə stɪk]	etw falsch verstehen
	cover up [ˌkʌvər 'ʌp]	abdecken
	fence off [ˌfens 'ɒf]	abschranken
	nearby ['nɪəbaɪ]	benachbart, in der Nähe
	make a note of [meɪk ə 'nəʊt əv]	notieren, beachten
	impossible [ɪm'pɒsəbl]	unmöglich
44	**installer** [ɪn'stɔ:lə]	Installateur/in
	guide [gaɪd]	Richtlinie, Einführung
	comply with [kəm'plaɪ wɪð]	befolgen
	satisfactory [ˌsætɪs'fæktəri]	zufrieden stellend, befriedigend
	running ['rʌnɪŋ]	laufend
	pressure ['preʃə]	Druck
	exclusive(ly) [ɪk'sklu:sɪv, -li]	ausschließlich, alleinig
	fused [fju:zd]	abgesichert
	capable of ['keɪpəbl əv]	fähig, in der Lage zu
	provide [prə'vaɪd]	liefern
	rated ['reɪtɪd]	Nenn-
	suitably rated ['su:təbli 'reɪtɪd]	mit geeigneten/passenden Nenndaten
	fit [fɪt]	installieren, einbauen
	stopcock ['stɒpkɒk]	Absperrhahn
	closure ['kləʊʒə]	(Ab-)Sperrung
	supplied [sə'plaɪd]	mitgeliefert
	shower outlet ['ʃaʊə aʊtlet]	Duschanschluss
	spray plate ['spreɪ pleɪt]	Düsenkopf
	become frozen [bɪˌkʌm 'frəʊzn]	einfrieren
45	**result from** [rɪ'zʌlt frəm]	resultieren aus, sich ergeben aus
	chest [tʃest]	Brust(korb)
	strength [streŋθ]	Kraft, Stärke

	amperage ['æmpərɪdʒ]	Amperezahl, Stromstärke
	duration [dju'reɪʃn]	Dauer
	painful ['peɪnfl]	schmerzhaft
	slight [slaɪt]	leicht
	tingle ['tɪŋgl]	Prickeln, Kribbeln
	unpleasant [ʌn'pleznt]	unangenehm
	muscle ['mʌsl]	Muskel
	freeze [fri:z]	erstarren
	let go of [ˌlet 'gəʊ əv]	loslassen
	adult ['ædʌlt]	Erwachsene/r
	freezing ['fri:zɪŋ]	erstarrend
	extreme(ly) [ɪk'stri:m, -li]	äußerst, außerordentlich
	disturb [dɪ'stɜ:b]	stören
	rhythm ['rɪðəm]	Rhythmus
	heartbeat ['hɑ:tbi:t]	Herzschlag
	ventricular fibrillation [venˌtrɪkjʊlə ˌfaɪbrɪ'leɪʃn]	Herzflimmern, Kammerflattern
	uncoordinated [ˌʌnkəʊ'ɔ:dɪneɪtɪd]	unkoordiniert
	heart [hɑ:t]	Herz
	nerve [nɜ:v]	Nerv
	damage ['dæmɪdʒ]	Schaden, Schäden
	likely ['laɪkli]	wahrscheinlich
	cardiac arrest [ˌkɑ:diæk ə'rest]	Herzstillstand
	serious ['sɪəriəs]	schlimm, schwer, ernst
	internal [ɪn'tɜ:nl]	inner(lich)
	external [ɪk'stɜ:nl]	äußer(lich)
	possible ['pɒsəbl]	möglich
	pain [peɪn]	Schmerz(en)
	mild [maɪld]	mild, leicht
	moderate ['mɒdərət]	mäßig
	severe [sɪ'vɪə]	schwer, stark
	first aid [ˌfɜ:st 'eɪd]	erste Hilfe

UNIT 5

46	**network** ['netwɜ:k]	Netz(werk), Sendenetz; vernetzen
	stand-alone (computer) [kəm'pju:tə(r)]	Einzel(computer)
	slow down [ˌsləʊ 'daʊn]	(sich) verlangsamen
	share [ʃeə]	(sich) teilen, gemeinsam (be)nutzen
	printer ['prɪntə]	Drucker
	internet connection [ˌɪntənet kə'nekʃn]	Internetanschluss, -zugang
	modem ['məʊdem]	Modem
	inefficient [ˌɪnɪ'fɪʃnt]	unwirtschaftlich
	back up [ˌbæk 'ʌp]	(auf Sicherheitskopien) speichern
	valuable ['væljuəbl]	wertvoll
	data ['deɪtə]	Daten
	time-consuming ['taɪm kənsju:mɪŋ]	zeitraubend
	realistic [ˌri:ə'lɪstɪk]	realistisch
	peripheral [pə'rɪfərəl]	Peripherie(gerät)
	hardware ['hɑ:dweə]	Geräte
	allow [ə'laʊ]	ermöglichen, gestatten
	secure [sɪ'kjuə]	sicher
	broadband internet connection [ˌbrɔ:dbænd 'ɪntənet kənekʃn]	Breitband-Internetanschluss, -zugang
	among [ə'mʌŋ]	zwischen, unter
	cost-effective [ˌkɒstɪ'fektɪv]	kostengünstig
	flexible ['fleksəbl]	flexibel
47	**be worried about** [bi 'wʌrid əbaʊt]	sich Sorgen machen wegen
	requirement [rɪ'kwaɪəmənt]	Erfordernis, Anforderung
	wireless ['waɪələs]	drahtlos, Funk-
	layout plan ['leɪaʊt plæn]	Grundriss
	sales department ['seɪlz dɪpɑ:tmənt]	Verkauf(sabteilung)
	concerned about [kən'sɜ:nd əbaʊt]	besorgt wegen
	WLAN (Wireless Local Area Network) [ˌdʌblju: 'læn, ˌwaɪələs 'netwɜ:k]	lokales drahtloses Funknetz
48	**fairly** ['feəli]	ziemlich, recht
	receptionist [rɪ'sepʃənɪst]	Empfang(smitarbeiter/in)
	Best wishes [best 'wɪʃɪz]	Viele Grüße, Mit freundlichen Grüßen
	meaning ['mi:nɪŋ]	Bedeutung
49	**topology** [tə'pɒlədʒi]	Topologie, Architektur
	option ['ɒpʃn]	Option, (Wahl-)Möglichkeit
	star [stɑ:]	Stern
	coaxial [kəʊ'æksiəl]	Koaxial-
	apply to [ə'plaɪ tə]	gelten/zutreffen für
	workstation ['wɜ:k steɪʃn]	Arbeitsplatz (mit Computer)
	manage ['mænɪdʒ]	umgehen (mit)
50	**configuration** [kənˌfɪgə'reɪʃn]	Konfiguration, Geräteanordnung
	suggestion [sə'dʒestʃən]	Vorschlag
	sensible ['sensəbl]	vernünftig
	reject [rɪ'dʒekt]	ablehnen
	clarification [ˌklærɪfɪ'keɪʃn]	Erklärung, Klärung, Erläuterung(en)
	call back [ˌkɔ:l 'bæk]	zurückrufen
51	**swap** [swɒp]	(ver)tauschen
	supply with [sə'plaɪ wɪð]	(be)liefern, zur Verfügung stellen
	up-to-date [ˌʌp tə 'deɪt]	aktuell, auf dem neuesten Stand
	ethernet ['i:θənet]	Ethernet
	cabling ['keɪblɪŋ]	Verkabelung
	security [sɪ'kjʊərəti]	Sicherheit
52	**chronological order** [krɒnəˌlɒdʒɪkl 'ɔ:də]	zeitliche Reihen-, Abfolge

	familiarize with [fə'mɪliəraɪz wɪð]	(sich) vertraut machen mit
	configure [kən'fɪgə]	konfigurieren
	operating system ['ɒpəreɪtɪŋ sɪstəm]	Betriebssystem
	supplier [sə'plaɪə]	Anbieter/in, Lieferant/in
	Kind regards [kaɪnd rɪ'gɑ:dz]	Mit freundlichen Grüßen
	All the best [ɔ:l ðə 'best]	Mit den besten Grüßen/ Wünschen
	hub [hʌb]	Verteiler(kasten)
	network card ['netwɜ:k kɑ:d]	Netzwerkkarte
	patchpanel ['pætʃpænl]	Patchpanel, Verteilerfeld
53	**rack** [ræk]	Gestell, Ständer
	slot [slɒt]	Schlitz
	ventilation [ˌventɪ'leɪʃn]	Lüftung, Belüftung
	assess [ə'ses]	beurteilen, begutachten
	facility [fə'sɪləti]	Vor-, Einrichtung
	surround [sə'raʊnd]	umgeben
	proper ['prɒpə]	ordnungsgemäß, angemessen
	mount bracket ['maʊnt brækɪt]	Montageträger
	chassis ['ʃæsi]	Grundplatte
	carrier tray ['kæriə treɪ]	CD-Schlitten
	careful(ly) ['keəfl, -fəli]	sorgfältig, vorsichtig
	slide [slaɪd]	schieben
	capture ['kæptʃə]	Halte-
	simultaneous(ly) [ˌsɪml'teɪniəs, -li]	gleichzeitig
	processor ['prəʊsesə]	Prozessor
	module ['mɒdju:l]	Modul
	tight(ly) ['taɪtli, -li]	fest
	power cord ['paʊə kɔ:d]	Netz-, Stromkabel
	crimp [krɪmp]	Schlaufe
	excessive [ɪk'sesɪv]	übermäßig
	wear [weə]	Abnutzung
	incomplete [ˌɪnkəm'pli:t]	unvollständig
	connection [kə'nekʃn]	Verbindung
	according to [ə'kɔ:dɪŋ tə]	entsprechend
54	**additional** [ə'dɪʃənl]	zusätzlich
	reasonable ['ri:znəbl]	angemessen, günstig
55	**towards** [tə'wɔ:dz]	in Richtung auf
	barrier ['bæriə]	Schranke, Barriere
	prevent [prɪ'vent]	(ver)hindern, verhüten
	unauthorized [ˌʌn'ɔ:θəraɪzd]	unbefugt, unberechtigt
	process ['prəʊses]	Prozess, Ablauf, Vorgang
	persuasive [pə'sweɪsɪv]	überzeugend
	advertising ['ædvətaɪzɪŋ]	Werbung, Reklame
	claim [kleɪm]	Behauptung
	hasty ['heɪsti]	hastig, überhastet
	impact ['ɪmpækt]	(Aus-)Wirkung
	resource [rɪ'sɔ:s]	Hilfsmittel, Ressource
	particular [pə'tɪkjələ]	besonder(s), speziell
	infrastructure ['ɪnfrəstrʌktʃə]	Infrastruktur
	unnecessary [ʌn'nesəsəri]	unnötig
	impractical [ɪm'præktɪkl]	unpraktisch, unbrauchbar
	implement ['ɪmplɪmənt]	einsetzen, installieren
	correct(ly) [kə'rekt, -li]	richtig
	gain [geɪn]	gewinnen, erlangen
	deploy [dɪ'plɔɪ]	einsetzen
	restrict [rɪ'strɪkt]	be-, einschränken
	frustrate ['frʌstreɪt]	frustrieren
	block [blɒk]	blockieren
	unavailable [ˌʌnə'veɪləbl]	nicht verfügbar
	in addition [ɪn ə'dɪʃn]	darüber hinaus, zudem
	responsibility [rɪˌspɒnsə'bɪləti]	Verantwortlichkeit
	common ['kɒmən]	häufig, weit verbreitet
	misconception [ˌmɪskən'sepʃn]	falsche Annahme
	protection [prə'tekʃn]	Schutz
	attack [ə'tæk]	Angriff
	percentage [pə'sentɪdʒ]	Prozentsatz
	incident ['ɪnsɪdənt]	Vorfall, Ereignis
	back door [ˌbæk 'dɔ:]	Hintertür
	typical(ly) ['tɪpɪkl, -kli]	normal(erweise)
	dial ['daɪəl]	wählen
	establish [ɪ'stæblɪʃ]	herstellen
	virus ['vaɪrəs]	Virus
	malicious [mə'lɪʃəs]	böswillig, mutwillig
	code [kəʊd]	Code
	contradict [ˌkɒntrə'dɪkt]	widersprechen

UNIT 6

56	**environment** [ɪn'vaɪrənmənt]	Umwelt
	conventional [kən'venʃənl]	konventionell
	alternative [ɔ:l'tɜ:nətɪv]	alternativ
	biomass ['baɪəmæs]	Biomasse
	geothermal [ˌdʒi:əʊ'θɜ:ml]	Erdwärme-
	hydro-electric [ˌhaɪdrəʊ ɪ'lektrɪk]	Wasserkraft-
	methane ['mi:θeɪn]	Methan
	nuclear ['nju:kliə]	Atom-
	tidal ['taɪdl]	Gezeiten-
	prefer [prɪ'fɜ:]	bevorzugen
	trade [treɪd]	Fach
	solar thermal collector [ˌsəʊlə ˌθɜ:ml kə'lektə]	Sonnen(wärme)kollektor
	PV (photovoltaic) cell [ˌpi: 'vi:, ˌfəʊtəʊvɒl'teɪɪk sel]	Photovoltaik-Zelle
57	**join** [dʒɔɪn]	sich anschließen, teilnehmen an
	abundant [ə'bʌndənt]	reichlich (vorhanden)

	non-polluting [ˌnɒnpəˈluːtɪŋ]	ohne Schadstoffe
	bill [bɪl]	Rechnung
	domestic [dəˈmestɪk]	häuslich, Haushalts-
	radiator [ˈreɪdieɪtə]	Heizkörper
	on average [ɒn ˈævərɪdʒ]	im Durchschnitt
	surface [ˈsɜːfɪs]	(Ober-)Fläche
	facing [ˈfeɪsɪŋ]	in Richtung
	exploit [ɪkˈsplɔɪt]	be-, ausnutzen
	greenhouse effect [ˈgriːnhaʊs ɪfekt]	Treibhauseffekt
	visible [ˈvɪzəbl]	sichtbar
	plate [pleɪt]	Platte
	insulated [ˈɪnsjuleɪtɪd]	isoliert
	collector [kəˈlektə]	Sammler/in, Kollektor
	housing [ˈhaʊzɪŋ]	Gehäuse, Verkleidung
	trap [træp]	einfangen
	reflect [rɪˈflekt]	reflektieren
	infrared [ˌɪnfrəˈred]	infrarot
	radiation [ˌreɪdiˈeɪʃn]	Strahlung
	consequence [ˈkɒnsɪkwəns]	Folge, Konsequenz
	fluid [ˈfluːɪd]	Flüssigkeit
	antifreeze [ˈæntifriːz]	Frostschutzmittel
	circulate [ˈsɜːkjəleɪt]	zirkulieren
	heat up [ˌhiːt ˈʌp]	erhitzen, erwärmen
	transfer [træns'fɜː]	weiterleiten
	reservoir [ˈrezəvwɑː]	Reservoir
	preserve [prɪˈzɜːv]	erhalten
	finite [ˈfaɪnaɪt]	begrenzt, endlich
	fossil fuel [ˈfɒsl fjuːəl]	fossiler Brennstoff
	coal [kəʊl]	Kohle
	oil [ɔɪl]	Öl
	natural gas [ˌnætʃrəl ˈgæs]	Erdgas
	noise [nɔɪz]	Geräusch, Lärm
	pollution [pəˈluːʃn]	Verschmutzung
	associated with [əˈsəʊʃieɪtɪd wɪð]	verbunden mit
	outage [ˈaʊtɪdʒ]	Ausfall
	recent [ˈriːsnt]	jüngst
	technological [ˌteknəˈlɒdʒɪkl]	technologisch
	breakthrough [ˈbreɪkθruː]	Durchbruch
	federal [ˈfedərəl]	Bundes-
	incentive [ɪnˈsentɪv]	Anreiz
	take into account [ˌteɪk ɪntu əˈkaʊnt]	in Betracht ziehen, berücksichtigen
	public [ˈpʌblɪk]	öffentlich
	utility [juːˈtɪləti]	Versorgungsbetrieb
	reliability [rɪˌlaɪəˈbɪləti]	Zuverlässigkeit
	durability [ˌdjʊərəˈbɪləti]	Dauer(haftigkeit)
	outstanding [aʊtˈstændɪŋ]	hervorragend
	principle [ˈprɪnsəpl]	Grundsatz, Prinzip
	silicon [ˈsɪlɪkən]	Silizium, Silikon
	sunlight [ˈsʌnlaɪt]	Sonnenlicht
	strike [straɪk]	aufschlagen
	certain [ˈsɜːtn]	gewiss, bestimmt
	portion [ˈpɔːʃn]	Portion, Anteil
	absorb [əbˈsɔːb]	absorbieren
	semiconductor [ˌsemikənˈdʌktə]	Halbleiter
	electron [ɪˈlektrɒn]	Elektron
	knock loose [ˌnɒk ˈluːs]	freisetzen
	force [fɔːs]	zwingen
	grid [grɪd]	Stromnetz
	future [ˈfjuːtʃə]	Zukunft
	thermal collector [ˌθɜːml kəˈlektə]	Wärmekollektor
	definitely [ˈdefɪnətli]	bestimmt, sicher(lich)
	deplete [dɪˈpliːt]	erschöpfen
	run out [ˌrʌn ˈaʊt]	zu Ende gehen
	meanwhile [ˈmiːnwaɪl]	inzwischen
	environmental [ɪnˌvaɪrənˈmentl]	Umwelt-
	steady (-ily) [ˈstedi, -ɪli]	stetig
58	**rely on** [rɪˈlaɪ ɒn]	angewiesen sein auf, sich stützen auf
	at present [ət ˈpreznt]	momentan, gegenwärtig
	come from [ˈkʌm frəm]	stammen von
	forever [fəˈrevə]	ewig
	contain [kənˈteɪn]	enthalten
	rarely [ˈreəli]	selten
	relevant [ˈreləvənt]	relevant, wichtig
	Danish [ˈdeɪnɪʃ]	dänisch
	specialist [ˈspeʃəlɪst]	Spezialist/in
	calculate [ˈkælkjuleɪt]	kalkulieren, berechnen
	renewable [rɪˈnjuːəbl]	erneuerbar
	efficiency [ɪˈfɪʃnsi]	Effizienz
59	**talk sb through sth** [ˌtɔːk ˈθruː]	jdm etw erläutern
	stand for [ˈstænd fə]	stehen für
	suitable for [ˈsuːtəbl fə]	geeignet für
	at first sight [ət fɜːst ˈsaɪt]	auf den ersten Blick
	roof [ruːf]	Dach
	orientation [ˌɔːriənˈteɪʃn]	Orientierung, Richtung
	inclination [ˌɪnklɪˈneɪʃn]	Neigung
	angle [ˈæŋgl]	Winkel
	degree [dɪˈgriː]	Grad
	enable [ɪˈneɪbl]	(es) ermöglichen
	all year round [ɔːl jɪə ˈraʊnd]	das ganze Jahr (hindurch)
	shade [ʃeɪd]	beschatten
	interconnect [ˌɪntəkəˈnekt]	miteinander verbunden sein
	array [əˈreɪ]	Ansammlung, Reihe
	refer to [rɪˈfɜː tə]	sich beziehen auf
	meteorological [ˌmiːtiərəˈlɒdʒɪkl]	meteorologisch
	rainfall [ˈreɪnfɔːl]	Niederschlag, Regen
	factor [ˈfæktə]	Faktor
	sense [sens]	Sinn
	consumption [kənˈsʌmpʃn]	Verbrauch

	environmentally friendly [ɪnˌvaɪrənˌmentəli 'frendli]	umweltfreundlich
	waste [weɪst]	verschwenden
60	**decimal** ['desɪml]	Dezimalzahl
	thousand ['θaʊznd]	tausend
	fraction ['frækʃn]	Bruch(teil)
60p	**sum** [sʌm]	Summe, Betrag
61	**worksheet** ['wɜːkʃiːt]	Arbeitsbogen
	determine [dɪ'tɜːmɪn]	bestimmen, feststellen
	method ['meθəd]	Methode, Verfahren
	multiply ['mʌltɪplaɪ]	multiplizieren
	divide [dɪ'vaɪd]	teilen, dividieren
	subtract [səb'trækt]	abziehen, subtrahieren
	insolation [ˌɪnsə'leɪʃn]	Sonnenbestrahlung
	inefficiency [ˌɪnɪ'fɪʃnsi]	Unzulänglichkeit
	peak [piːk]	Gipfel, Spitze, Höchstwert
	wattage ['wɒtɪdʒ]	Wattleistung
	round up to [ˌraʊnd 'ʌp tə]	aufrunden
62	**grid-tied** ['grɪd taɪd]	an das Stromnetz angeschlossen
	label ['leɪbl]	beschriften
	schematic [skiː'mætɪk]	schematisch(e Darstellung), Grafik
62t	**agree** [ə'griː]	übereinstimmen, sich einigen
	independent [ˌɪndɪ'pendənt]	unabhängig
	entirely [ɪn'taɪəli]	ganz, vollständig
	battery ['bætəri]	Batterie
	charge controller ['tʃɑːdʒ kəntrəʊlə]	Ladungssteuergerät
	charge [tʃɑːdʒ]	Ladung
	overcharge [ˌəʊvə'tʃɑːdʒ]	überladen
	provider [prə'vaɪdə]	Anbieter, Versorger
	meter ['miːtə]	Zähler
	contribution [ˌkɒntrɪ'bjuːʃn]	Beitrag, Lieferung
63	**planet** ['plænɪt]	Planet
	global warming [ˌgləʊbl 'wɔːmɪŋ]	Erderwärmung, Erwärmung der Erdatmosphäre
	government ['gʌvənmənt]	Regierung
	argue ['ɑːgjuː]	behaupten, argumentieren
	wind turbine ['wɪnd tɜːbaɪn]	Windturbine, -rad
	initiative [ɪ'nɪʃətɪv]	Initiative
	hydrogen ['haɪdrədʒən]	Wasserstoff
	FAQs [ˌef eɪ 'kjuːz]	häufig gestellte Fragen
	tense [tens]	Zeit(form)
	sell back to [ˌsel 'bæk tə]	verkaufen an
	owner ['əʊnə]	Besitzer/in, Eigentümer/in
	emission [ɪ'mɪʃn]	Emission
64p	**in series** [ɪn 'sɪəriːz]	in Reihe (geschaltet)
	explanation [ˌeksplə'neɪʃn]	Erklärung
64	**controller** [kən'trəʊlə]	Steuergerät
	operating manual [ˌɒpəreɪtɪŋ 'mænjuəl]	Handbuch, Bedienungsanleitung
	moisture ['mɔɪstʃə]	Feuchtigkeit
	sleep mode [sliːp məʊd]	Bereitschaft, Ruhezustand
	shutdown ['ʃʌtdaʊn]	Abschaltung
	personnel [ˌpɜːsə'nel]	Personal, Mitarbeiter
	feature ['fiːtʃə]	Merkmal
	enclosure [ɪn'kləʊʒə]	Umhüllung, Ummantelung
	exposed [ɪk'spəʊzd]	freiliegend
	remain [rɪ'meɪn]	bleiben
	closed [kləʊzd]	geschlossen
	except [ɪk'sept]	außer
	servicing instruction [ˌsɜːvɪsɪŋ ɪn'strʌkʃn]	(Betriebs-)Anleitung
	risk [rɪsk]	Risiko
	specify ['spesɪfaɪ]	aufführen, angeben
	operating instruction [ˌɒpəreɪtɪŋ ɪn'strʌkʃn]	(Bedienungs-)Anleitung
	unless [ən'les]	wenn nicht, es sei denn
	cabinet ['kæbɪnət]	Schaltschrank
	dew [djuː]	Tau
	front panel indicators [ˌfrʌnt pænl 'ɪndɪkeɪtəz]	Anzeigen der Frontplatte
	incorporate [ɪn'kɔːpəreɪt]	umfassen
	coloured ['kʌləd]	farbig
	LED (light emitting diode) indicator [laɪt i'mɪtɪŋ'daɪəʊd']	LED (Leuchtdiode)n-Anzeige
	mode [məʊd]	Modus
	abnormal [æb'nɔːml]	anormal, ungewöhnlich
	reset [ˌriː'set]	zurückstellen, zurücksetzen (technisch)
	extend [ɪk'stend]	verlängern, ausziehen
	sufficient [sə'fɪʃnt]	ausreichend
	front door [ˌfrʌnt 'dɔː]	Vordertür
	depress [dɪ'pres]	(nieder)drücken
	initiate [ɪ'nɪʃieɪt]	einleiten
65	**measure up** [ˌmeʒər 'ʌp]	den Erwartungen entsprechen
	eco-friendly ['iːkə frendli]	umweltfreundlich
	retreat [rɪ'triːt]	Rückzug, Zufluchtsort
	meditation [ˌmedɪ'teɪʃn]	Meditation
	emit [ɪ'mɪt]	emittieren, ausstoßen
	offset ['ɒfset]	ausgleichen
	plant [plɑːnt]	pflanzen
	dirty ['dɜːti]	schmutzig
	dry cleaners [ˌdraɪ 'kliːnəz]	chemische Reinigung
	redesign [ˌriːdɪ'zaɪn]	neu entwerfen, umstrukturieren

	shrub [ʃrʌb]	Busch, Strauch
	nappy ['næpi]	Windel
	disposable [dɪ'spəʊzəbl]	Einweg-, Wegwerf-
	laundry ['lɔ:ndri]	Wäsche, Wasch-
	takeaway [ˌteɪkə'weɪ]	zum Mitnehmen
	polystyrene cup [ˌpɒli'staɪri:n kʌp]	Plastiktasse
	tidy (up) [ˌtaɪdi 'ʌp]	aufräumen
	go back and forth [gəʊ ˌbæk ənd 'fɔ:θ]	hin und her gehen
	turn off [ˌtɜ:n 'ɒf]	ausschalten
	tin [tɪn]	Dose, Büchse
	hotplate ['hɒtpleɪt]	Koch-, Warmhalteplatte
	pan [pæn]	Topf, Pfanne
	diameter [daɪ'æmɪtə]	Durchmesser
	significant [sɪg'nɪfɪkənt]	bedeutend, spürbar
	reduction [rɪ'dʌkʃn]	Verringerung, Senkung
	rather than ['rɑ:ðə ðən]	anstatt
	tumble dryer [ˌtʌmbl 'draɪə]	Wäschetrockner

UNIT 7

66	**identify** [aɪ'dentɪfaɪ]	identifizieren, festlegen
	capacitor [kə'pæsɪtə]	Kondensator
	diode ['daɪəʊd]	Diode
	full-wave bridge rectifier [ˌfʊl weɪv ˌbrɪdʒ 'rektɪfaɪə]	Zweiwegbrückengleichrichter
	integrated voltage stabilizer [ˌɪntɪgreɪtɪd ˌvəʊltɪdʒ 'steɪbəlaɪzə]	integrierter Spannungsstabilisierer
	mains input socket [ˌmeɪnz ˌɪnpʊt 'sɒkɪt]	Stromverteilerdose, Netzanschlussdose
	resistor [rɪ'zɪstə]	Widerstand
	transistor [træn'zɪstə]	Transistor
	valve [vælv]	Ventil
67	**ham radio operator** [hæm ˌreɪdiəʊ 'ɒpəreɪtə]	Amateurfunker/in
	issue ['ɪʃu:]	Ausgabe
	radio enthusiast [ˌreɪdiəʊ ɪn'θju:ziæst]	begeisterter Amateurfunker
	sensory check ['sensəri tʃek]	Sicht-, Geruchsprüfung
	famous ['feɪməs]	berühmt
	amplifier ['æmplɪfaɪə]	Verstärker
	blow [bləʊ]	durchbrennen
	standby ['stændbaɪ]	Bereitschaft, Ruhezustand
	illuminate [ɪ'lu:mɪneɪt]	beleuchten
	input ['ɪnpʊt]	Eingang (technisch)
	remove [rɪ'mu:v]	entfernen, abnehmen
	charred [tʃɑ:d]	verkohlt
	luckily ['lʌkɪli]	glücklicherweise, zum Glück
	output ['aʊtpʊt]	Ausgang (technisch)
	realize ['rɪəlaɪz]	sich klar werden/sein
	take back [ˌteɪk 'bæk]	mitnehmen
68	**trace** [treɪs]	finden, zurückverfolgen
69	**take turns** [teɪk 'tɜ:nz]	(sich) abwechseln
	square [skweə]	Quadrat, Kästchen
70t	**(guitar) amp** [gɪ'tɑ:r æmp]	(Gitarren-)Verstärker
	GFCI (ground fault circuit interrupter) (AE = RCD) [ˌdʒi: ef si: 'aɪ, ˌgraʊnd fɔ:lt ˌsɜ:kɪt ˌɪntə'rʌptə, ˌɑ: si: 'di:]	Fehlerstromschutzschalter
	short circuit [ˌʃɔ:t 'sɜ:kɪt]	Kurzschluss
70	**paper clip** ['peɪpə klɪp]	Büroklammer
	solder joint ['səʊldə dʒɔɪnt]	Lötstelle
	circuit board ['sɜ:kɪt bɔ:d]	(Schaltungs)Platine
	soldered ['səʊldəd]	gelötet
	come loose [kʌm 'lu:s]	sich lösen
71	**browse (through)** ['braʊz θru:]	durchsuchen, (herum)stöbern (in)
	post [pəʊst]	verschicken
	rectifier ['rektɪfaɪə]	Gleichrichter
	throw away [ˌθrəʊ ə'weɪ]	wegwerfen
	appreciate [ə'pri:ʃieɪt]	dankbar sein (für)
	cheers [tʃɪəz]	tschüs, danke
	look up [ˌlʊk 'ʌp]	heraussuchen
	forward ['fɔ:wəd]	vorwärts
	reverse [rɪ'vɜ:s]	Rück-, entgegengesetzt
	solder in [ˌsəʊldər 'ɪn]	anlöten
	power supply ['paʊə səplaɪ]	Stromanschluss
	plug in [ˌplʌg 'ɪn]	(Stecker) einstecken
	obvious ['ɒbviəs]	offensichtlich, sichtbar
	smoothing circuit [ˌsmu:ðɪŋ 'sɜ:kɪt]	Glättungsschaltung
	discoloured [dɪs'kʌləd]	verfärbt
	leading ['li:dɪŋ]	führend
	base [beɪs]	Platine, Fuß, Basis
	unsolder [ˌʌn'səʊldə]	entlöten
	desoldering pump [ˌdi:'səʊldərɪŋ pʌmp]	Entlöt(saug)pumpe
	suck off [ˌsʌk 'ɒf]	absaugen
	set [set]	(ein)stellen
	discover [dɪ'skʌvə]	entdecken, feststellen
	consequently ['kɒnsɪkwəntli]	infolgedessen
	defective [dɪ'fektɪv]	fehlerhaft
	reconnect [ˌri:kə'nekt]	wieder anschließen
72	**battery-operated** ['bætəri ɒpəreɪtɪd]	batteriebetrieben
	take sth. apart [teɪk ə'pɑ:t]	etw auseinandernehmen
	mounted ['maʊntɪd]	montiert
	bring out [ˌbrɪŋ 'aʊt]	herausbringen
	come out [ˌkʌm 'aʊt]	erscheinen, herauskommen

	track [træk]	Stück, Song
	unplugged [ˌʌn'plʌgd]	unplugged, ohne Verstärker
	sock it to s.o. ['sɒk ɪt tə]	es jdm zeigen
	lyrics ['lɪrɪks]	(Lied-)Text
	guy [gaɪ]	Kerl, Typ
	dedicate ['dedɪkeɪt]	widmen
74	**chat** [tʃæt]	Chat, Unterhaltung
	recording studio [rɪ'kɔ:dɪŋ stju:diəʊ]	Aufnahmestudio
	special ['speʃl]	Spezial(anfertigung)
	take care [teɪk 'keə]	pass auf dich auf!
	release [rɪ'li:s]	herausbringen, veröffentlichen
	record ['rekɔ:d]	(Schall-)Platte
	examine [ɪg'zæmɪn]	untersuchen, prüfen
	screw up [ˌskru: 'ʌp]	zusammenschrauben
	shorten ['ʃɔ:tn]	(ver)kürzen
	unplug [ˌʌn'plʌg]	den Stecker herausziehen
75	**tube (AE = valve BE)** [tju:b, vælv]	Röhre
	high-power RF application [ˌhaɪ paʊər ˌɑ:r 'ef ˌæplɪ'keɪʃn]	Anwendung, Einsatz im Hochleistungs-Radiofrequenzbereich
	radio station ['reɪdiəʊ steɪʃn]	Rundfunkstation, Radiosender
	power tube ['paʊə tju:b]	(Sende)röhre, Röhre mit hoher Leistung
	frequency ['fri:kwənsi]	Frequenz
	broadcast station ['brɔ:dkɑ:st steɪʃn]	Sender
	powered by ['paʊəd baɪ]	betrieben mit
	transmitter [træns'mɪtə]	Sender
	economic [ˌi:kə'nɒmɪk]	wirtschaftlich
	worldwide ['wɜ:ldwaɪd]	weltweit
	hybrid ['haɪbrɪd]	hybrid
	make up [ˌmeɪk 'ʌp]	ausmachen
	tube amplifier [ˌtju:b 'æmplɪfaɪə]	Röhrenverstärker
	tone [təʊn]	Ton
	distortion [dɪ'stɔ:ʃn]	Verzerrung
	characteristic [ˌkærəktə'rɪstɪk]	(typisches) Merkmal, Eigenschaft
	unique [ju'ni:k]	einzigartig, einmalig
	simulate ['sɪmjuleɪt]	simulieren
	insist on [ɪn'sɪst ɒn]	bestehen auf
	vacuum tube technology [ˌvækjuəm tju:b tek'nɒlədʒi]	Röhrentechnologie
	abbreviation [əˌbri:vi'eɪʃn]	Abkürzung
	oscillation [ˌɒsɪ'leɪʃn]	Schwingung
	wave [weɪv]	Welle
	range [reɪndʒ]	Bereich, Spektrum
	wavelength ['weɪvleŋθ]	Wellenlänge
	vary ['veəri]	verändern, variieren
	communicate [kə'mju:nɪkeɪt]	übermitteln
	approx. (approximately) [ə'prɒks, ə'prɒksɪmətli]	ungefähr, etwa, zirka

UNIT 8

76	**ammeter** ['æmi:tə]	Amperemeter, Strom(stärke)messer
	capacitance meter [kə'pæsɪtəns mi:tə]	Kapazitätsmesser
	mains tester ['meinz testə]	Spannungsprüfer
	ohmmeter ['əʊmmi:tə]	Ohmmeter, Widerstandsmesser
	oscilloscope [ə'sɪləskəʊp]	Oszilloskop
	test light ['test laɪt]	Prüfleuchte
	voltmeter ['vəʊltmi:tə]	Spannungsmessgerät, -messer, Voltmeter
	measurement ['meʒəmənt]	Maß
76p	**read out** [ˌri:d 'aʊt]	(laut) vorlesen
77	**RMS (root mean square) voltage** [ˌɑ:r em 'es, ˌru:t mi:n skweə 'vəʊltɪdʒ]	Effektivspannung
	peak voltage ['pi:k vəʊltɪdʒ]	Spitzenspannung
	square root ['skweə ru:t]	Quadratwurzel
77p	**calculator** ['kælkjuleɪtə]	(Taschen-)Rechner
	work out [ˌwɜ:k 'aʊt]	ausrechnen
78	**graph** [grɑ:f]	Kurve, Diagramm
	screen [skri:n]	Bildschirm
	vertical (Y) axis [ˌvɜ:tɪkl 'waɪ æksɪs]	vertikale (Y-) Achse
	display [dɪ'spleɪ]	(an)zeigen
	waveform ['weɪvfɔ:m]	Wellenform
	alternate ['ɔ:ltɜ:neɪt]	abwechseln
	V/t graph [ˌvi: 'ti: grɑ:f]	Spannungs-Zeit-Diagramm
	horizontal (X) axis [ˌhɒrɪzɒntl 'eks æksɪs]	horizontale (X-)Achse
	division [dɪ'vɪʒn]	Abteilung
	knob [nɒb]	Knopf
	adjust [ə'dʒʌst]	einstellen, regeln
	intensity [ɪn'tensəti]	Intensität
	brightness ['braɪtnəs]	Helligkeit
	focus ['fəʊkəs]	scharf einstellen, bündeln
	sinusoidal [ˌsaɪnə'sɔɪdl]	sinusförmig
	rectified ['rektɪfaɪd]	gleichgerichtet
79	**battery charger** ['bætəri tʃɑ:dʒə]	Batterieladegerät, Akku-Lader
	exchange student [ɪks'tʃeɪndʒ stju:dnt]	Austauschstudent/in

	stage [steɪdʒ]	Phase, Stadium
79t	**regulator** ['regjuleɪtə]	Regler
	adjustable [ə'dʒʌstəbl]	verstellbar
	recharge [ˌriː'tʃɑːdʒ]	aufladen
	rechargeable [ˌriː'tʃɑːdʒəbl]	aufladbar
79	**double-check** ['dʌbl tʃek]	(noch einmal) nachprüfen
	confirmation [ˌkɒnfə'meɪʃn]	Bestätigung
80	**electrolytic capacitor** [ɪlektrəˌlɪtɪk kə'pæsɪtə]	Elektrolytkondensator
	writing ['raɪtɪŋ]	Schrift
	integrated ['ɪntɪgreɪtɪd]	integriert
	colour-coded ['kʌlə kəʊdɪd]	farbig markiert
	potentiometer [pəˌtenʃi'ɒmɪtə]	Potentiometer
81	**limit** ['lɪmɪt]	begrenzen
	decrease [dɪ'kriːs]	verringern
	conduct away [kən'dʌkt əweɪ]	ableiten, abführen
	define [dɪ'faɪn]	definieren, (genauer) erklären
82	**write up** [ˌraɪt 'ʌp]	aufschreiben
	experiment [ɪk'sperɪmənt]	Versuch, Experiment
	figure ['fɪgə]	Zahl, Ziffer
	assignment [ə'saɪnmənt]	Arbeit, Aufgabe
	circuit diagram ['sɜːkɪt daɪəgræm]	Schaltbild
	attachment [ə'tætʃmənt]	Anlage
	step by step [ˌstep baɪ 'step]	Schritt für Schritt
	pulsate [pʌl'seɪt]	pulsieren
	drop [drɒp]	Rückgang
	smooth [smuːð]	glätten
	ripple ['rɪpl]	Welligkeit (technisch)
	eliminate [ɪ'lɪmɪneɪt]	eliminieren, beseitigen
	stabilizer ['steɪbəlaɪzə]	Stabilisator
83	**previous** ['priːviəs]	vorhergehend, vorherig
	smoothing capacitor [ˌsmuːðɪŋ kə'pæsɪtə]	Sieb-, Glättungskondensator
	integrated voltage regulator [ˌɪntɪgreɪtɪd ˌvəʊltɪdʒ 'regjuleɪtə]	integrierter Spannungsregler
84	**terrible** ['terəbl]	schrecklich, furchtbar
	go crazy [gəʊ 'kreɪzi]	wütend werden, ausrasten
	phone card ['fəʊn kɑːd]	Telefonkarte
	any longer [eni 'lɒŋgə]	länger
85	**in excess of** [ɪn 'ekses əv]	über (hinaus)
	floating operation [ˌfləʊtɪŋ ˌɒpə'reɪʃn]	erdfreier Betrieb
	series ['sɪəriːz]	Reihe
	limiting ['lɪmɪtɪŋ]	Begrenzung
	SOA (Safe Operation Area) control [ˌes əʊ 'eɪ seɪf ˌɒpə'reɪʃn 'earia kən'trəʊl]	Regelung des sicheren Arbeitsbereiches
	monolithic [ˌmɒnə'lɪθɪk]	monolithisch
	be intended for [bi ɪn'tendɪd fə]	dienen zu
	by means of [baɪ 'miːnz əv]	mit Hilfe von, durch
	resistive divider [rɪˌzɪstɪv dɪ'vaɪdə]	Widerstands-spannungsteiler
	exceptional(ly) [ɪk'sepʃənl, -ʃənəli]	außergewöhnlich
	sheet [ʃiːt]	Blatt (Papier)
	desired [dɪ'zaɪəd]	gewünscht
	voltage divider [ˌvəʊltɪdʒ dɪ'vaɪdə]	Spannungsteiler

UNIT 9

86	**breakdown** ['breɪkdaʊn]	Panne, Störung
87	**spin** [spɪn]	schleudern, (schnell) drehen
	timer control ['taɪmə kəntrəʊl]	Zeitschalter
	be worthwhile [bi ˌwɜːθ'waɪl]	sich lohnen
	awkward ['ɔːkwəd]	unangenehm, lästig
88	**stupid(ly)** ['stjuːpɪd, -li]	dumm, blöde
	development [dɪ'veləpmənt]	Entwicklung
	neither ['naɪðə]	auch nicht, keine
	frequent ['friːkwənt]	häufig, oft
	advisable [əd'vaɪzəbl]	ratsam
	economical [ˌiːkə'nɒmɪkl]	sparsam, wirtschaftlich
	apparently [ə'pærəntli]	offensichtlich, anscheinend
	unusual [ʌn'juːʒuəl]	ungewöhnlich
	tub [tʌb]	(Wasch-)Trommel
	drain [dreɪn]	entleeren
	speed [spiːd]	Geschwindigkeit
	expertise [ˌekspɜː'tiːz]	Fach-, Sachkenntnis
	clutch [klʌtʃ]	Kupplung
	engaged [ɪn'geɪdʒd]	eingelegt
	brake [breɪk]	Bremse
	disengaged [ˌdɪsɪn'geɪdʒd]	nicht betätigt
	wash motor ['wɒʃ məʊtə]	Waschmotor
	in forward gear [ɪn 'fɔːwəd gɪə]	im Vorwärtsgang
	in reverse gear [ɪn rɪ'vɜːs gɪə]	im Rückwärtsgang
89	**corresponding to** [ˌkɒrɪ'spɒndɪŋ tə]	entsprechend

	indicate ['ɪndɪkeɪt]	(an)zeigen
	reach [riːtʃ]	erreichen
	out of adjustment [aʊt əv ə'dʒʌstmənt]	falsch eingestellt
	cylinder ['sɪlɪndə]	Zylinder
	stuck [stʌk]	fest-, eingeklemmt
	disk brake ['dɪsk breɪk]	Scheibenbremse
	energize ['enədʒaɪz]	aufladen
	air pressure ['eə preʃə]	Luftdruck
	leak [liːk]	undichte Stelle, Leck
	exhaust [ɪg'zɔːst]	Auspuff
	motor starter ['məʊtə stɑːtə]	Startermotor
	heater coil ['hiːtə kɔɪl]	Heizspirale
	adjustment [ə'dʒʌstmənt]	Einstellung
	escape from [ɪ'skeɪp frəm]	ausfließen aus
	container [kən'teɪnə]	Behälter, Gefäß
89p	**domestic appliance technician** [də,mestɪk ə,plaɪəns tek'nɪʃn]	Hausgerätetechniker/in
	call out [,kɔːl 'aʊt]	ausrufen
	manual ['mænjuəl]	Handbuch
90	**specification** [,spesɪfɪ'keɪʃn]	Angabe
	enquiry [ɪn'kwaɪəri]	Anfrage
	conclude [kən'kluːd]	(ab)schließen
91	**freestanding** [,friː'stændɪŋ]	freistehend
	performance [pə'fɔːməns]	Leistung(en)
	effectiveness [ɪ'fektɪvnəs]	Effektivität
	gradation [grə'deɪʃn]	Einteilung
	at a glance [ət ə 'glɑːns]	auf einen Blick
	shining example [,ʃaɪnɪŋ ɪg'zɑːmpl]	leuchtendes Beispiel
	represent [,reprɪ'zent]	darstellen
	scale [skeɪl]	Skala
	based upon ['beɪst əpɒn]	auf der Grundlage
	in accordance with [ɪn ə'kɔːdəns wɪð]	entsprechend, gemäß
	standardized ['stændədaɪzd]	standardisiert, genormt
	procedure [prə'siːdʒə]	Verfahren
	cotton ['kɒtn]	Baumwolle
	clamp [klæmp]	Klammer
	motor shaft ['məʊtə ʃɑːft]	Motorwelle
	tie [taɪ]	befestigen, binden
	transmission [træns'mɪʃn]	Triebwerk, Getriebe
	motor coupling ['məʊtə kʌplɪŋ]	Motorkupplung
	pump [pʌmp]	Pumpe
	harness ['hɑːnɪs]	Verkleidung

	hold [həʊld]	(fest)halten
	mounting clamp ['maʊntɪŋ klæmp]	Montageklammer
	thumb [θʌm]	Daumen
	detach [dɪ'tætʃ]	abnehmen
	transmission shaft [træns'mɪʃn ʃɑːft]	Antriebs-, Getriebewelle
	rubber union ['rʌbə juːniən]	Gummiverbindung
	prong [prɒŋ]	Zinke, Zacke, Spitze, Dorn
92	**armature** ['ɑːmətʃə]	Armatur; Anker
	commutator ['kɒmjuteɪtə]	Stromwender
	download [,daun'ləʊd]	herunterladen
	placement machine [,pleɪsmənt mə'ʃiːn]	Bestückungsmaschine
	repel [rɪ'pel]	(sich) abstoßen
	rotor ['rəʊtə]	Rotor, Läufer
	stator ['steɪtə]	Stator, Ständer, feststehender Teil
93	**scientific** [,saɪən'tɪfɪk]	wissenschaftlich
	like poles ['laɪk pəʊlz]	gleiche Pole
	unlike poles ['ʌnlaɪk pəʊlz]	ungleiche Pole
	attract [ə'trækt]	(sich) anziehen
	electromagnet [ɪ,lektrəʊ'mægnɪt]	Elektromagnet
	transform [træns'fɔːm]	um-, verwandeln
	opposing [ə'pəʊzɪŋ]	entgegengesetzt
	interaction [,ɪntər'ækʃn]	Interaktion
	exciter field [ɪk'saɪtə fiːld]	Erregerfeld
	torque [tɔːk]	Drehmoment, -kraft
	split-ring commutator [splɪt ,rɪŋ 'kɒmjuteɪtə]	Kommutator, Stromwender (mit Stromwenderlamellen)
	brush against [brʌʃ ə'genst]	bürsten an
	reversal [rɪ'vɜːsl]	Umkehr(ung)
	interact with [,ɪntər'ækt wɪð]	sich gegenseitig beeinflussen
	alternately [ɔːl'tɜːnətli]	abwechselnd
	shaft [ʃɑːft]	Achse, Welle
	alternator ['ɔːltəneɪtə]	Wechselstromgenerator
	high-speed ['haɪspiːd]	Hochgeschwindigkeits-
	cycle ['saɪkl]	Zyklus, Drehung
	exceed [ɪk'siːd]	übersteigen, -schreiten
	rotation [rəʊ'teɪʃn]	Drehung
	blender ['blendə]	Mischer, Mixer
	vacuum cleaner ['vækjuəm kliːnə]	Staubsauger
	hair dryer ['heə draɪə]	Föhn
94	**continuity** [,kɒntɪ'njuːəti]	(Strom-)Versorgung
	power rating ['paʊə reɪtɪŋ]	Stromverbrauch
	probe [prəʊb]	Sonde
	power wire ['paʊə waɪə]	(flexibles) Stromkabel
	in turn [ɪn 'tɜːn]	der Reihe nach

	service technician [ˌsɜ:vɪs tek'nɪʃn]	Wartungstechniker/in
	patient ['peɪʃnt]	geduldig
	complicated ['kɒmplɪkeɪtɪd]	kompliziert
95	**joke** [dʒəʊk]	Witze machen
	spokesman ['spəʊksmən]	Sprecher
	swear [sweə]	schwören
	purchaser ['pɜ:tʃəsə]	Käufer/in
	complain [kəm'pleɪn]	(sich) beschweren
	slip out of [ˌslɪp 'aʊt əv]	herausrutschen
	cup holder ['kʌp həʊldə]	Tassenhalter
	tray [treɪ]	Schlitten
	sophisticated [sə'fɪstɪkeɪtɪd]	hoch entwickelt, kompliziert
	digital camera [ˌdɪdʒɪtl 'kæmərə]	Digitalkamera
	advanced [əd'vɑ:nst]	fortgeschritten
	versatile ['vɜ:sətaɪl]	vielseitig
	cell phone (AE = mobile phone BE) ['sel fəʊn, ˌməʊbaɪl 'fəʊn]	Handy, Mobiltelefon
	handheld organizer [ˌhændheld 'ɔ:gənaɪzə]	Taschencomputer
	refuse [rɪ'fju:z]	(es) ablehnen, sich weigern
	marketing ['mɑ:kɪtɪŋ]	Werbe-, Vertriebs-, Marketing(-)
	customer service [ˌkʌstəmə 'sɜ:vɪs]	Kundendienst
	consultant [kən'sʌltənt]	Berater/in
	corporate executive [ˌkɔ:pərət ɪg'zekjətɪv]	leitende/r Angestellte/r
	instruction manual [ɪn'strʌkʃn mænjuəl]	Bedienungsanleitung, Benutzerhandbuch
	retailer ['ri:teɪlə]	Einzelhändler/in
	perfect(ly) ['pɜ:fɪkt, -li]	vollkommen
	lid [lɪd]	Deckel
	ink cartridge [ɪŋk 'kɑ:trɪdʒ]	Tintenpatrone
	dramatic [drə'mætɪk]	dramatisch
	complexity [kəm'pleksəti]	Komplexität
	willing ['wɪlɪŋ]	gewillt
	read through [ˌri:d 'θru:]	durchlesen
	producer [prə'dju:sə]	Produzent/in, Hersteller/in
	user manual [ju:zə 'mænjuəl]	Benutzerhandbuch
	come with ['kʌm wɪð]	geliefert werden mit
	digital display [ˌdɪdʒɪtl dɪ'spleɪ]	Digitalanzeige
	china ['tʃaɪnə]	Porzellan
	crystal ['krɪstl]	Kristall
	volume ['vɒlju:m]	Volumen, Menge, Umfang
	filter out [ˌfɪltər 'aʊt]	herausfiltern
	hard minerals [hɑ:d 'mɪnərəlz]	feste Mineralien
	abridged (from) [ə'brɪdʒd frəm]	gekürzt (aus)
	aspect ['æspekt]	Aspekt, Seite
	guideline ['gaɪdlaɪn]	Richtlinie, Hinweis

UNIT 10

96	**cartoon** [kɑ:'tu:n]	Karikatur
	caption ['kæpʃn]	Bildunterschrift
	cartoonist [kɑ:'tu:nɪst]	Karikaturist/in
	standardization [ˌstændədaɪ'zeɪʃn]	Standardisierung
	competence ['kɒmpɪtəns]	Kompetenz
	achievement [ə'tʃi:vmənt]	Leistung, Errungenschaft
	IEC (International Electrotechnical Commission) [ˌaɪ i: 'si:, ɪntəˌnæʃnəl ɪˌlektrəʊˌteknɪkl kə'mɪʃn]	Internationale Elektrotechnische Kommission
97	**property** ['prɒpəti]	Eigentum
	improve [ɪm'pru:v]	verbessern
	trading ['treɪdɪŋ]	Geschäfts-, Handels-
	doorway ['dɔ:weɪ]	(Haus-)Eingang, Tür
	take place [teɪk 'pleɪs]	stattfinden
	chaotic [keɪ'ɒtɪk]	chaotisch
	gauge [geɪdʒ]	Spurweite
	locomotive [ˌləʊkə'məʊtɪv]	Lok(omotive)
	in mid-journey [ɪn ˌmɪd 'dʒɜ:ni]	mitten auf der Reise
	predate [ˌpri:'deɪt]	zeitlich vorangehen
	ISO (International Organization for Standardization) [ˌaɪ es 'əʊ, ɪntəˌnæʃnəl ˌɔgənaɪˌzeɪʃn fə ˌstændədaɪ'zeɪʃn]	Internationale Organisation für Normung
	take for granted [ˌteɪk fə 'grɑ:ntɪd]	als selbstverständlich betrachten/ansehen
	found [faʊnd]	gründen
	consensus [kən'sensəs]	Konsens, Einigung
	voluntary ['vɒləntri]	freiwillig
	workable ['wɜ:kəbl]	anwendbar
	compatibility [kəmˌpætə'bɪləti]	Kompatibilität
	rigid ['rɪdʒɪd]	streng, fest
	directive [də'rektɪv]	Richtlinie, Anordnung
	committee [kə'mɪti]	Ausschuss
	climate ['klaɪmət]	Klima
	developing [dɪ'veləpɪŋ]	Entwicklungs-
	industrialized [ɪn'dʌstriəlaɪzd]	Industrie-
	structure ['strʌktʃə]	Struktur
	overseeing [ˌəʊvə'si:ɪŋ]	Aufsichts-

	publish ['pʌblɪʃ]	veröffentlichen, herausgeben
	harmonize ['hɑ:mənaɪz]	harmonisieren, abstimmen
	convergence [kən'vɜ:dʒəns]	Konvergenz, Zusammenpassen
	HD television [ˌeɪtʃ 'di: 'telɪvɪʒn]	HD-Fernsehen
	multimedia [ˌmʌlti'mi:diə]	Multimedia-
	ITU (International Telecommunications Union) [ˌaɪ ti: 'ju:, ɪntəˌnæʃnəl ˌtelikəˌmju:nɪˌkeɪʃnz 'ju:niən]	Internationale Fernmeldeunion
	jungle ['dʒʌŋgl]	Dschungel
	productive(ly) [prə'dʌktɪv, -li]	produktiv
	PAN (Personal Area Network) [ˌpi: eɪ 'en, ˌpɜ:sənl ˌeəriə 'netwɜ:k]	Kleinstnetzwerk zur persönlichen Kommunikation
	Wi-Fi [ˌwaɪ 'faɪ]	Wi-Fi
	vital ['vaɪtl]	wesentlich, unerlässlich, unbedingt notwendig
	despite [dɪ'spaɪt]	trotz
	challenge ['tʃælɪndʒ]	Herausforderung, Aufgabe
	increasing(ly) [ɪn'kri:sɪŋ, -li]	zunehmend
	dependent [dɪ'pendənt]	abhängig
	engine ['endʒɪn]	Motor
98	**agreement** [ə'gri:mənt]	bereinstimmung
	willing(ly) ['wɪlɪŋ, -li]	bereit(willig)
	ordinary ['ɔ:dnri]	gewöhnlich, normal
	weak [wi:k]	schwach
	inflexible [ɪn'fleksəbl]	unflexibel
	interfere [ˌɪntə'fɪə]	sich einmischen, stören
	topic ['tɒpɪk]	Thema
	brief [bri:f]	kurz
	importance [ɪm'pɔ:tns]	Wichtigkeit, Bedeutung
	telecommunications [ˌtelikəˌmju:nɪ'keɪʃnz]	Fernmeldewesen
	motor vehicle [ˌməʊtə 'vi:əkl]	Kraftfahrzeug
99	**programmer** ['prəʊgræmə]	Programmierer/in
	serve [sɜ:v]	bedienen
100	**disparity** [dɪ'spærəti]	Ungleichheit
	adopt [ə'dɒpt]	an-, übernehmen
	pin [pɪn]	Stift, Nadel
	apply for [ə'plaɪ fə]	sich bewerben um
	pointed ['pɔɪntɪd]	spitz
	take over [ˌteɪk 'əʊvə]	übernehmen
101	**blow up** [ˌbləʊ 'ʌp]	explodieren
	throughout [θru:'aʊt]	überall (in)
	differ ['dɪfə]	sich unterscheiden
	seeming(ly) ['si:mɪŋ, -li]	anscheinend
	firework display [ˌfaɪəwɜ:k dɪ'spleɪ]	Feuerwerk
	so-called [ˌsəʊ 'kɔ:ld]	so genannt
	grounding contact [ˌgraʊndɪŋ 'kɒntækt]	Erdkontakt
	ungrounded [ˌʌn'graʊndɪd]	nicht geerdet
	parallel ['pærəlel]	parallel
	in line [ɪn 'laɪn]	in (einer) Reihe
	rectangular [rek'tæŋgjələ]	rechteckig
	triangle ['traɪæŋgl]	Dreieck
	grounding pin ['graʊndɪŋ pɪn]	Erdstift
	arrange [ə'reɪndʒ]	anordnen
	popular(ly) ['pɒpjələ, -li]	allgemein
	offset ['ɒfset]	nicht in einer Reihe
102	**consider** [kən'sɪdə]	bedenken, berücksichtigen
	checklist ['tʃek lɪst]	Checkliste
	rank [ræŋk]	einstufen, anordnen
	handout ['hændaʊt]	Merkblatt
	take into consideration [ˌteɪk ɪntə kənˌsɪdə'reɪʃn]	berücksichtigen
	customs ['kʌstəmz]	Zoll
103	**column** ['kɒləm]	Spalte
	keyword ['ki:wɜ:d]	Stichwort
	attitude ['ætɪtju:d]	Einstellung, Haltung
	body language ['bɒdi læŋgwɪdʒ]	Körpersprache
103t	**raise** [reɪz]	wecken
	cross-cultural [ˌkrɒs'kʌltʃərəl]	multikulturell
	awareness [ə'weənəs]	Bewusstsein
	do business with [du 'bɪznəs wɪð]	Geschäfte machen mit
	aware of [ə'weər əv]	bewusst
	similarity [ˌsɪmə'lærəti]	hnlichkeit
	effective(ly) [ɪ'fektɪv, -li]	wirkungsvoll
	misinterpretation [ˌmɪsɪnˌtɜ:prɪ'teɪʃn]	Missverständnis
	pattern ['pætn]	Muster
	illustrate ['ɪləstreɪt]	illustrieren, veranschaulichen
	on time [ɒn 'taɪm]	pünktlich
	successful [sək'sesfl]	erfolgreich
	respect [rɪ'spekt]	(be)achten, respektieren
	alien ['eɪliən]	fremd
	relationship [rɪ'leɪʃnʃɪp]	Beziehung, Verbindung
	as far as … is concerned [əz ˌfɑ:r əz ɪz kən'sɜ:nd]	was … anbetrifft
	hierarchy ['haɪərɑ:ki]	Hierarchie
	status ['steɪtəs]	Status, (soziale) Stellung
	highest-ranking [ˌhaɪɪst 'ræŋkɪŋ]	höchstrangig
	executive [ɪg'zekjətɪv]	leitende/r Angestellte/r, Manager/in

	personal contact [ˌpɜːsənl ˈkɒntækt]	körperliche Berührung
	at all cost [ət ɔːl ˈkɒst]	unter allen Umständen
	besides [bɪˈsaɪdz]	neben
	bowing [ˈbaʊɪŋ]	Verbeugen, Verneigen
	nod [nɒd]	nicken
	handshaking [ˈhændʃeɪkɪŋ]	Händeschütteln
	customary [ˈkʌstəməri]	üblich
	greeting [ˈgriːtɪŋ]	Begrüßung
	gentle [ˈdʒentl]	sanft, leicht
	essential [ɪˈsenʃl]	(absolut) notwendig
	appointment [əˈpɔɪntmənt]	Termin, Verabredung
	get-it-done [ˌget ɪt ˈdʌn]	ergebnisorientiert
	contrary to [ˈkɒntrəri tə]	gegenüber, im Gegensatz zu
	distributed [dɪˈstrɪbjuːtɪd]	verteilt
	decision making [dɪˈsɪʒn meɪkɪŋ]	Entscheidungsfindung
	touch-oriented [ˌtʌtʃ ˈɔːrientɪd]	kontaktorientiert
	apart [əˈpɑːt]	entfernt
	personal comfort zone [ˈpɜːsənl ˈkʌmfət zəʊn]	Diskretionsabstand
	atomic clock [əˌtɒmɪk ˈklɒk]	Atomuhr
	innumerable [ɪˈnjuːmərəbl]	unzählig, zahllos
	institute [ˈɪnstɪtjuːt]	Institut
	conform to [kənˈfɔːm tə]	entsprechen
	generalize [ˈdʒenrəlaɪz]	verallgemeinern
	bear in mind [ˌbeər ɪn ˈmaɪnd]	beachten
	behaviour [bɪˈheɪvjə]	Verhalten, Benehmen
	stereotype [ˈsteriətaɪp]	Klischee(vorstellung)
	non-productive [ˌnɒn prəˈdʌktɪv]	unproduktiv
	strategy [ˈstrætədʒi]	Strategie
	observe [əbˈzɜːv]	beobachten, beachten
103	**move on to** [ˌmuːv ˈɒn tə]	übergehen zu
	feel free [fiːl ˈfriː]	sich nicht scheuen
	interrupt [ˌɪntəˈrʌpt]	unterbrechen
	talk about [ˈtɔːk əbaʊt]	sprechen über
	bar chart [ˈbɑː tʃɑːt]	Säulen-, Balkendiagramm
	key point [ˈkiː pɔɪnt]	Kernpunkt, -frage
	trade fair [ˈtreɪd feə]	Handelsmesse
104	**fair** [feə]	Messe
	extensive [ɪkˈstensɪv]	umfassend, umfangreich
	database [ˈdeɪtəbeɪs]	Datenbank
	organizer [ˈɔːgənaɪzə]	Organisator/in, Veranstalter/in
	exhibitor [ɪgˈzɪbɪtə]	Aussteller/in
	inquire about [ɪnˈkwaɪər əbaʊt]	sich erkundigen nach
	re-write [ˌriː ˈraɪt]	umschreiben, neu schreiben
	representative [ˌreprɪˈzentətɪv]	Vertreter/in, Repräsentant/in
	launch [lɔːntʃ]	einführen, herausbringen
	brochure [ˈbrəʊʃə]	Broschüre
	leaflet [ˈliːflət]	Prospekt
	detailed [ˈdiːteɪld]	detailliert, ausführlich
	business card [ˈbɪznəs kɑːd]	(Visiten-)Karte
105	**pitfall** [ˈpɪtfɔːl]	Falle, Fallstrick
	custom [ˈkʌstəm]	Sitte, Brauch
	failure [ˈfeɪljə]	Fehlschlag, Versagen
	ignorance [ˈɪgnərəns]	Unwissenheit, Unkenntnis
	client [ˈklaɪənt]	Kunde/Kundin
	index finger [ˈɪndeks fɪŋgə]	Zeigefinger
	circle [ˈsɜːkl]	Kreis
	vulgar [ˈvʌlgə]	vulgär
	offend [əˈfend]	beleidigen
	get down to business [get daʊn tə ˈbɪznəs]	zur Sache kommen, an die Arbeit gehen
	Arab [ˈærəb]	arabisch
	character [ˈkærəktə]	Charakter, Figur, Person

UNIT 11

106	**textile mill** [ˈtekstaɪl mɪl]	Textilfabrik
	brewery [ˈbruːəri]	Brauerei
	thermostat [ˈθɜːməstæt]	Thermostat
	refrigeration [rɪˌfrɪdʒəˈreɪʃn]	Kühlung
	feeder [ˈfiːdə]	Zubringer, Einzug
	coolant flow monitor [ˌkuːlənt fləʊ ˈmɒnɪtə]	Kühlmittelkontrollanzeige
	robotic [rəʊˈbɒtɪk]	Roboter-
	welding [ˈweldɪŋ]	Schweißen, Schweiß-
	weaving machine [ˈwiːvɪŋ məˌʃiːn]	Webmaschine
	brewing [ˈbruːɪŋ]	Brau(erei)-
	machinery [məˈʃiːnəri]	Maschinen(park), Geräte
	agenda [əˈdʒendə]	Tagesordnung
	minutes [ˈmɪnɪts]	Protokoll
107	**venue** [ˈvenjuː]	(Tagungs-)Ort
	chair [tʃeə]	Leiter/in
	project manager [ˈprɒdʒekt mænɪdʒə]	Projektmanager/in
	attendee [əˌtenˈdiː]	Anwesende/r
	objective [əbˈdʒektɪv]	Ziel, Zweck
	review [rɪˈvjuː]	Besprechung, berblick
	apology [əˈpɒlədʒi]	Entschuldigung
	preparation [ˌprepəˈreɪʃn]	Vorbereitung
	subcontractor [ˌsʌbkənˈtræktə]	Subunternehmer
	monitoring [ˈmɒnɪtərɪŋ]	Überwachung, Kontrolle
	supervision [ˌsuːpəˈvɪʒn]	Aufsicht, Überwachung
	commissioning [kəˈmɪʃnɪŋ]	Inbetriebnahme
	acceptance [əkˈseptəns]	Über-, Annahme

	on site [ɒn 'saɪt]	auf der Baustelle, vor Ort
	construction [kən'strʌkʃn]	Bau, Konstruktion
	head office [ˌhed 'ɒfɪs]	Zentrale
	report [rɪ'pɔ:t]	berichten
	progress ['prəʊgres]	Fortschritt/e
	to date [tə 'deɪt]	bisher
	confirm [kən'fɜ:m]	bestätigen
	alongside [əˌlɒŋ'saɪd]	neben
	at the latest [ət ðə 'leɪtɪst]	spätestens
	remind [rɪ'maɪnd]	(jdn an etw) erinnern
	on a weekly basis [ɒn ə ˌwi:kli 'beɪsɪs]	wöchentlich
	event [ɪ'vent]	Ereignis, Vorgang
108	**written** ['rɪtn]	schriftlich
	arrangement [ə'reɪndʒmənt]	Vorbereitung
	leave for ['li:v fə]	abreisen nach
	depart from [dɪ'pɑ:t frəm]	abfliegen, starten
	stopover ['stɒpəʊvə]	Zwischenlandung
	airline ['eəlaɪn]	Fluggesellschaft
	fly with ['flaɪ wɪð]	fliegen mit
	driving licence ['draɪvɪŋ laɪsns]	Führerschein
109	**collect** [kə'lekt]	abholen
	supervise ['su:pəvaɪz]	beaufsichtigen
	be set for [bi 'set fə]	eingestellt sein auf
	sort out [ˌsɔ:t 'aʊt]	klären, lösen
	let sb know [ˌlet 'nəʊ]	jdm Bescheid geben/ sagen, jdn informieren
	rep (representative) [rep,reprɪ'zentativ]	Vertreter/in
110	**passenger** ['pæsɪndʒə]	Passagier/in, Fluggast
	check-in desk ['tʃekɪn desk]	Abfertigungsschalter
	proceed to [prə'si:d tə]	sich begeben nach
	gate [geɪt]	Flugsteig, Ausgang
	board [bɔ:d]	an Bord gehen
	delayed [dɪ'leɪd]	verspätet
	consult [kən'sʌlt]	beachten
	departure [dɪ'pɑ:tʃə]	Abflug
	unattended [ˌʌnə'tendɪd]	unbeaufsichtigt
	check-in counter ['tʃek ɪn kaʊntə]	Abfertigungsschalter
	declare [dɪ'kleə]	verzollen, zur Verzollung anmelden
	luggage ['lʌgɪdʒ]	Gepäck
	check in [ˌtʃek 'ɪn]	abgeben
	purpose ['pɜ:pəs]	Zweck
	hand luggage ['hænd lʌgɪdʒ]	Handgepäck
	belt [belt]	Band
	en route [ˌɒn 'ru:t]	unterwegs
	information desk [ɪnfə'meɪʃn desk]	Informationsschalter, Auskunft
	over weight [ˌəʊvə 'weɪt]	Übergewicht
	security control [sɪ'kjʊərəti kəntrəʊl]	Sicherheitskontrolle
	metal detector [ˌmetl dɪ'tektə]	Metallsuchgerät
	go off [ˌgəʊ 'ɒf]	losgehen, ansprechen
	customs counter ['kʌstəmz kaʊntə]	Zollschalter
110t	**clerk** [klɑ:k]	Angestellte/r
	shuttle train ['ʃʌtl treɪn]	Zubringerzug
	airport official ['eəpɔ:t əfɪʃl]	Flughafenangestellte/r
	pack [pæk]	(ein)packen
	carry ['kæri]	befördern, bei sich haben/führen
	anyone else [ˌeniwʌn 'els]	jemand anderer
	firearm ['faɪərɑ:m]	Schusswaffe
	explosive [ɪk'spləʊsɪv]	Sprengstoff
	boarding pass ['bɔ:dɪŋ pɑ:s]	Bordkarte
	window seat ['wɪndəʊ si:t]	Fensterplatz
	taken ['teɪkən]	besetzt
	never mind [ˌnevə 'maɪnd]	macht nichts
	step through ['step θru:]	durchgehen
	customs declaration card [ˌkʌstəmz ˌdeklə'reɪʃn kɑ:d]	Zoll(inhalts)erklärung
	workmate ['wɜ:kmeɪt]	(Arbeits-)Kollege
	beverage ['bevərɪdʒ]	Getränk
	foodstuff ['fu:dstʌf]	Nahrungsmittel
111	**setting** ['setɪŋ]	Einstellung
	specially designed [ˌspeʃəli dɪ'zaɪnd]	speziell entwickelt
	impressive [ɪm'presɪv]	beeindruckend, eindrucksvoll
	flow meter ['fləʊ mi:tə]	Durchflussregler
	RPM (revolutions per minute) [ˌɑ: pi: 'em, revəˌlu:ʃnz pə 'mɪnɪt]	(Um-)Drehungen pro Minute
	instruct [ɪn'strʌkt]	einweisen, instruieren
	transfer ['trænsfɜ:]	Übermittlung
	policy ['pɒləsi]	(Firmen-)Politik
	come up [ˌkʌm 'ʌp]	sich ergeben
	handover ['hændəʊvə]	Übergabe
	under guidance [ˌʌndə 'gaɪdəns]	unter Leitung
	satisfied with ['sætɪsfaɪd wɪð]	zufrieden mit
112	**brewmaster** ['bru:mɑ:stə]	Braumeister/in
	memory ['meməri]	Speicher
	sensor ['sensə]	Sensor, (Mess-)Fühler
	temperature indicator [ˌtemprətʃə 'ɪndɪkeɪtə]	Temperaturanzeige
	automated ['ɔ:təmeɪtɪd]	automatisiert
	relay ['ri:leɪ]	Relais

	automotive [ˌɔ:tə'məʊtɪv]	Auto(mobil)-
	programmable ['prəʊgræməbl]	programmierbar
	sequence ['si:kwəns]	Reihenfolge
	timing ['taɪmɪŋ]	Zeitplanung
	analogue ['ænəlɒg]	analog
	automation [ˌɔ:tə'meɪʃn]	Automation
	pneumatic [nju:'mætɪk]	pneumatisch
	hydraulic [haɪ'drɔ:lɪk]	hydraulisch
	assembly line [ə'sembli laɪn]	Montage-, Fließband
	interface ['ɪntəfeɪs]	Schnittstelle
	attached to [ə'tætʃt tə]	verbunden mit, angeschlossen an
	proprietary [prə'praɪətri]	Eigentümer-, Besitzer-
	cam timer ['kæm ˌtaɪmə]	Nockensteuerung
	initially [ɪ'nɪʃəli]	anfänglich, zuerst
	revision [rɪ'vɪʒn]	Änderung, Überarbeitung
	hardwired ['hɑ:dwaɪəd]	fest verdrahtet
113	**audience** ['ɔ:diəns]	Publikum, Zuhörer
	in conclusion [ɪn kən'klu:ʒn]	zum Schluss, abschließend
114	**junior** ['dʒu:niə]	untergeordnet
	senior ['si:niə]	vorgesetzt, leitend
	confident ['kɒnfɪdənt]	sicher
	mediate ['mi:dieɪt]	vermitteln, dolmetschen
	pre-installation [ˌpri: ˌɪnstə'leɪʃn]	Vorinstallation
	in case of [ɪn 'keɪs əv]	im Falle von
	be glad [bi 'glæd]	sich freuen
	cooperate [kəʊ'ɒpəreɪt]	zusammenarbeiten
115	**appear** [ə'pɪə]	erscheinen
	forceful(ly) ['fɔ:sfl, -fəli]	kräftig
	stare into ['steər ɪntə]	anstarren
	uncomfortable [ʌn'kʌmftəbl]	unwohl
	peer [pɪə]	Kollege/-in
	lose face [lu:z 'feɪs]	das Gesicht verlieren
	instant(ly) ['ɪnstənt, -li]	sofort, umgehend
	destroy [dɪ'strɔɪ]	zerstören
	insult [ɪn'sʌlt]	beleidigen
	criticize ['krɪtɪsaɪz]	kritisieren
	make fun of [meɪk 'fʌn əv]	sich lustig machen über
	seated ['si:tɪd]	platziert
	descending [dɪ'sendɪŋ]	absteigend
	host [həʊst]	Gastgeber/in
	deliver [dɪ'lɪvə]	vortragen
	turn the floor over to [tɜ:n ðə ˌflɔ:r 'əʊvə tə]	das Wort übergeben an
	member ['membə]	Mitglied
	spokesperson ['spəʊkspɜ:sn]	Sprecher/in
	impolite [ˌɪmpə'laɪt]	unhöflich
	respond [rɪ'spɒnd]	antworten
	counterpart ['kaʊntəpɑ:t]	Gegenüber

	enthusiastic(ally) [ɪnˌθju:zi'æstɪk, -kli]	begeistert, enthusiastisch
	accidental(ly) [ˌæksɪ'dentl, -təli]	zufällig, unbeabsichtigt
	critical ['krɪtɪkl]	kritisch
	watch out for [ˌwɒtʃ 'aʊt fə]	aufpassen auf
	sense of humour [ˌsens əv 'hju:mə]	Sinn für Humor

UNIT 12

116	**electronic technician** [ɪlekˌtrɒnɪk tek'nɪʃn]	Elektroniker/in
	earn [ɜ:n]	verdienen
	salary ['sæləri]	Gehalt, Lohn
117	**register** ['redʒɪstə]	registrieren, eintragen lassen
	CV (curriculum vitae) [kəˌrɪkjələm 'vi:taɪ]	Lebenslauf
	recruitment agency [rɪ'kru:tmənt eɪdʒənsi]	Arbeits-, Stellenvermittlung
	switchboard ['swɪtʃbɔ:d]	Schalttafel
	wirer ['waɪərə]	Kabelleger/in
	negotiable [nɪ'gəʊʃiəbl]	Verhandlungssache
	reference ['refrəns]	Akten-, Bezugszeichen
	licensed ['laɪsnst]	lizensiert
	electrical fitter [ɪˌlektrɪkl 'fɪtə]	Elektroinstallateur/in
	rewarding [rɪ'wɔ:dɪŋ]	lohnend, befriedigend
	generous ['dʒenərəs]	großzügig
	plenty of ['plenti əv]	viel, reichlich
	attractive [ə'træktɪv]	attraktiv, reizvoll
	upload [ˌʌp'ləʊd]	senden
118	**summary** ['sʌməri]	Zusammenfassung
	work experience ['wɜ:k ɪkspɪəriəns]	Berufserfahrung
	extra-curricular activity [ˌekstrə kəˌrɪkjulə æk'tɪvəti]	außerschulische Aktivität
	fluent ['flu:ənt]	fließend
119	**personal details** [ˌpɜ:sənl 'di:teɪlz]	persönliche Angaben
	gender ['dʒendə]	Geschlecht
	marital status [ˌmærɪtl 'steɪtəs]	Familienstand
	seek [si:k]	suchen
	vocational [vəʊ'keɪʃənl]	Berufs-
	leaving certificate [ˌli:vɪŋ sə'tɪfɪkət]	Abschlusszeugnis
	primary education [ˌpraɪməri edʒu'keɪʃn]	Grundschulbildung
	secondary education [ˌsekəndri edʒu'keɪʃn]	höhere Schul(aus)bildung
	energy efficient [ˌenədʒi ɪ'fɪʃnt]	energiesparend
	familiar with [fə'mɪliə wɪð]	vertraut mit

	word processing ['wɜ:d prəʊsesɪŋ]	Textverarbeitung
	spreadsheet ['spredʃi:t]	Tabellenprogramm, -kalkulation
	certificate [sə'tɪfɪkət]	Urkunde
	life-saving ['laɪf seɪvɪŋ]	Lebensrettungs-
	rescue ['reskju:]	Rettung(s-)
	captain ['kæptɪn]	Kapitän
	soccer ['sɒkə]	Fußball
	referee [ˌrefə'ri:]	Referenz
120	**electrician's mate** [ɪˌlek'trɪʃnz meɪt]	Elektrotechnikgehilfe, -in
	webmaster ['webmɑ:stə]	Webmaster
	choir ['kwaɪə]	Chor
	construction site [kən'strʌkʃn saɪt]	Baustelle
	premises ['premɪsɪz]	Geschäft(sräume)
	higher education entrance qualification [ˌhaɪər edʒu'keɪʃn ˌentrəns kwɒlɪfɪ'keɪʃn]	Fachhochschulreife
	goods vehicle licence [gʊdz ˌvi:əkl 'laɪsns]	Lkw-Führerschein
	military service ['mɪlətri sɜ:vɪs]	Militärdienst
	traineeship [treɪ'ni:ʃɪp]	Ausbildung
	school leaving certificate [ˌsku:l li:vɪŋ sə'tɪfɪkət]	Mittlere Reife
	community service [kəˌmju:nəti'sɜ:vɪs]	Zivildienst
	vocational college (GB), vocational school (US) [vəʊˌkeɪʃənl 'kɒlɪdʒ, vəʊˌkeɪʃənl 'sku:l]	Berufsschule, -kolleg, gewerbliche Schule
	German university entrance qualification [ˌdʒɜ:mən ju:nɪ'vɜ:səti ˌentrəns kwɒlɪfɪ'keɪʃn]	Hochschulreife (Abitur)
	German higher education entrance qualification [ˌdʒɜ:mən ˌhaɪər edʒu'keɪʃn ˌentrəns ˌkwɒlɪfɪ'keɪʃn]	Fachhochschulreife
121	**cover letter** ['kʌvə letə]	Begleitschreiben
	vacancy ['veɪkənsi]	offene Stelle
	Ref. (reference) [ref, 'refrəns]	Akten-, Bezugszeichen, Betreffzeile
	above-mentioned [əˌbʌv 'menʃnd]	oben erwähnt
	theoretical [ˌθɪə'retɪkl]	theoretisch
	range from … to ['reɪndʒ frəm tə]	sich von … bis erstrecken
	operating theatre ['ɒpəreɪtɪŋ θɪətə]	Operationssaal
	building site ['bɪldɪŋ saɪt]	Baustelle
	enclose [ɪn'kləʊz]	beifügen
	at short notice [ət ˌʃɔ:t 'nəʊtɪs]	kurzfristig
	shortly ['ʃɔ:tli]	in Kürze
	command [kə'mɑ:nd]	Beherrschung
	grateful ['greɪtfl]	dankbar
	Enc. (enclosed) [ɪn'kləʊzd]	Anlage
122	**maintenance personnel** [ˌmeɪntənəns ˌpɜ:sə'nel]	Wartungspersonal
	electrical plant [ɪˌlektrɪkl 'plɑ:nt]	Elektroanlage
	offshore [ˌɒf'ʃɔ:]	vor der Küste
	petrochemical [ˌpetrəʊ'kemɪkl]	petrochemisch, Erdöl-
	recognize ['rekəgnaɪz]	anerkennen
	expiry date [ɪk'spaɪəri deɪt]	Ablaufdatum
	quote [kwəʊt]	angeben, nennen
	offline ['ɒflaɪn]	offline, nicht übers Internet
123	**robot** ['rəʊbɒt]	Roboter
	grade [greɪd]	Note
	punctual ['pʌŋktʃuəl]	pünktlich
123t	**take a seat** [ˌteɪk ə 'si:t]	Platz nehmen
	target ['tɑ:gɪt]	Ziel
	in the fresh air [ɪn ðə freʃ 'eə]	an der frischen Luft, im Freien
	agency ['eɪdʒənsi]	Agentur, Büro
	relief [rɪ'li:f]	Hilfe
	weakness ['wi:knəs]	Schwäche
123	**interviewer** ['ɪntəvju:ə]	Leiter/in des Vorstellungsgespräches, Interviewer/in
	Federal Agency for Technical Relief [ˌfedərəl ˌeɪdʒənsi fə ˌteknɪkl rɪ'li:f]	Technisches Hilfswerk
124p	**various** ['veəriəs]	verschiedene
	electrical engineering [ɪˌlektrɪkl ˌendʒɪ'nɪərɪŋ]	Elektrotechnik
	so far [ˌsəʊ 'fɑ:]	bis jetzt
	work placement [wɜ:k 'pleɪsmənt]	Praktikum
	tape recorder ['teɪp rɪkɔ:də]	Tonbandgerät
	canoeing club [kə'nu:ɪŋ klʌb]	Kanuverein
	canoe [kə'nu:]	Kanu fahren
	fetch [fetʃ]	holen, bringen
	clean up [ˌkli:n 'ʌp]	sauber machen, aufräumen
124	**final grade** ['faɪnl greɪd]	Abschlussnote

take on [ˌteɪk 'ɒn] einstellen
Arabic ['ærəbɪk] arabisch

125 **price tag** ['praɪs tæg] Preisschild
neck [nek] Hals
trendy ['trendi] modisch, schick
shiny ['ʃaɪni] glänzend
tights [taɪts] Strumpfhosen
evening wear ['iːvnɪŋ weə] Abendgarderobe
shave [ʃeɪv] (sich) rasieren
(well) in advance [wel ɪn əd'vɑːns] (rechtzeitig) vorher
settle ['setl] sich beruhigen
at the very least [ət ðə ˌveri 'liːst] zumindest
jewellery ['dʒuːəlri] Schmuck
insecurity [ˌɪnsɪ'kjʊərəti] Unsicherheit
chunky ['tʃʌŋki] dick, klobig
necklace ['nekləs] (Hals-)Kette
bracelet ['breɪslət] Armband
strong-smelling [ˌstrɒŋ 'smelɪŋ] stark riechend
garlic ['gɑːlɪk] Knoblauch
give in [ˌgɪv 'ɪn] abgeben
coat [kəʊt] Mantel
re-polish [ˌriː 'pɒlɪʃ] nochmals putzen
thorough(ly) ['θʌrə, -li] gründlich
nail [neɪl] (Finger-)Nagel
be with sb in a moment [bi wɪð ˌsʌmbədi ɪn ə 'məʊmənt] gleich bei jdm sein

A–Z WORD LIST

Dieses Wörterverzeichnis enthält alle Wörter aus *Electricity Matters* in alphabetischer Reihenfolge. Nicht aufgeführt sind die Wörter aus der Liste des Grundwortschatzes. t = das Wort befindet sich in den *Transcripts* (Hörverständnisübungen). p = das Wort befindet sich in den *Pairwork Files*.

A

abbreviation 75 Abkürzung
abnormal 64 anormal, ungewöhnlich
above-mentioned 121 oben erwähnt
abridged (from) 95 gekürzt (aus)
absolutely 19 absolut, völlig
absorb 57 absorbieren
abundant 57 reichlich (vorhanden)
accept 9 an-, übernehmen
acceptance 107 Über-, Annahme
accident 34 Unfall
accidental(ly) 115 zufällig, unbeabsichtigt
in accordance with 91 entsprechend, gemäß
according to 53 entsprechend
accounts 11 Buchhaltung(sunterlagen), Rechnungswesen
achievement 96 Leistung, Errungenschaft
act as 40 fungieren als
act out 14 vorspielen, aufführen
in addition 55 darüber hinaus, zudem
additional 54 zusätzlich
adjust 78 einstellen, regeln
adjustable 79t verstellbar
adjustment 89 Einstellung
administration 13 Verwaltung
administrative 11 Verwaltungs-
adopt 100 an-, übernehmen
adult 45 Erwachsene/r
(well) in advance 125 (rechtzeitig) vorher
advanced 95 fortgeschritten
advancement 15 Aufstieg
advantage 41 Vorteil
advertising 55 Werbung, Reklame
advice 37 Rat(schlag)
advisable 88 ratsam
advise 18 beraten
aerial 14 Antenne
afford 10 (es) sich leisten (können)
after-sales service 10 Kundendienst, Wartung
agency 123t Agentur, Büro
agenda 106 Tagesordnung
agree 62t übereinstimmen, sich einigen
agree to disagree 19 sich darauf einigen, dass man sich nicht einigen kann
agree with 19 zustimmen
agreement 98 Übereinstimmung
air conditioning 17 Klimaanlage
in the fresh air 123t an der frischen Luft, im Freien
in the open air 17 im Freien, an der frischen Luft
air pressure 89 Luftdruck
airline 108 Fluggesellschaft
airport official 110t Flughafenangestellte/r
airtight 8 luftdicht
alien 103t fremd
All the best 52 Mit den besten Grüßen/Wünschen
all year round 59 das ganze Jahr (hindurch)
Allen keys 20 Inbusschlüssel, Innensechskantschlüssel
allow 46 ermöglichen, gestatten
be not allowed 27 nicht dürfen
alongside 107 neben
alternate 78 abwechseln
alternately 93 abwechselnd
alternative 56 alternativ
alternatively 25 als Alternative, als weitere Möglichkeit
alternator 93 Wechselstromgenerator
ammeter 76 Amperemeter, Strom(stärke)messer
among 46 zwischen, unter
amount 10 Summe, Betrag
amp(ere) 40 Ampere
amperage 45 Amperezahl, Stromstärke
amp(lifier) 67 Verstärker
analogue 112 analog
angle 59 Winkel
angry 50 böse, wütend, verärgert
annoyed 39 verärgert, irritiert
antifreeze 57 Frostschutzmittel
any longer 84 länger
anyone else 110t jemand anderer
apart 103t entfernt
apology 107 Entschuldigung
apparently 88 offensichtlich, anscheinend
appear 115 erscheinen
appliance 27 Gerät
apply (for) 15 sich bewerben (um)
apply to 49 gelten/zutreffen für
appointment 103t Termin, Verabredung
appreciate 71 dankbar sein (für)
apprentice 25 Lehrling, Auszubildende/r
apprenticeship 22 Lehre, Ausbildung
appropriate 32 angemessen, passend
approx. (approximately) 75 ungefähr, etwa, zirka
Arab 105 arabisch
Arabic 124 arabisch
architect 7 Architekt/in
argue 63 behaupten, argumentieren
argument 19 Argument
armature winding resistance 23 Ankerwicklungswiderstand
armature 92 Armatur; Anker

armchair 31 Sessel
arrange 7, 101 anordnen
arrangement 108 Vorbereitung
array 59 Ansammlung, Reihe
as far as … is concerned 103t was … anbetrifft
as if 23 als ob, als wenn
aspect 95 Aspekt, Seite
assemble 16 zusammensetzen, -bauen, montieren
assembly line 112 Montage-, Fließband
assess 53 beurteilen, begutachten
assignment 82 Arbeit, Aufgabe
assistant 11 Assistent/in
associated with 57 verbunden mit
atomic clock 103t Atomuhr
attach 27 anschließen
attached to 112 verbunden mit, angeschlossen an
attachment 82 Anlage
attack 55 Angriff
attendee 107 Anwesende/r
attitude 103 Einstellung, Haltung
attract 93 (sich) anziehen
attractive 117 attraktiv, reizvoll
audience 113 Publikum, Zuhörer
automated 112 automatisiert
automation 112 Automation
automotive 112 Auto(mobil)-
available 17 anwesend, verfügbar
on average 57 im Durchschnitt
avionics 15 Avionik, Flugzeugelektronik
avoid 50 (ver)meiden
aware of 103t bewusst
awareness 103t Bewusstsein
awful 27 furchtbar, schrecklich
awkward 87 unangenehm, lästig

B

back door 55 Hintertür
back up 46 (auf Sicherheitskopien) speichern
balance 25 Verhältnis
bar chart 103 Säulen-, Balkendiagramm
barrier 55 Schranke, Barriere
base 71 Platine, Fuß, Basis
based in 12 mit (Haupt-)Sitz in
based upon 91 auf der Grundlage
basic 25 Grund-
basically 40 grundsätzlich
battery 62t Batterie
battery charger 79 Batterieladegerät, Akku-Lader
battery-operated 72 batteriebetrieben
bear in mind 103t beachten
behaviour 103t Verhalten, Benehmen
bell 31 Klingel, Glocke
belt 110 Band
bend 17 biegen
benefit 15 Unterstützung
besides 103t neben
Best wishes 48 Viele Grüße, Mit freundlichen Grüßen
beverage 110t Getränk
bill 57 Rechnung
bimetallic strip 41 Bimetallstreifen
biomass 56 Biomasse
bleeper 12 (Telefon-)Pieper
blender 93 Mischer, Mixer
block 55 blockieren
block of flats 17 (Wohn-)Block, Hochhaus
block release 25 Blockunterricht
blow 67 durchbrennen
blow up 101 explodieren
board 110 an Bord gehen
boarding pass 110t Bordkarte
body language 103 Körpersprache
bolt 21 Riegel, Bolzen, Schraube
bother 43 stören, auf die Nerven gehen
bottom 41 Unterseite
bowing 103t Verbeugen, Verneigen
bracelet 125 Armband
bracket 18 Klammer
brainstorm 19 (Ideen) sammeln
brake 88 Bremse
break 23 (unter)brechen
break down 10 kaputt gehen
breakdown 86 Panne, Störung
breakthrough 57 Durchbruch
brewery 106 Brauerei
brewing 106 Brau(erei)-
brewmaster 112 Braumeister/in
brief 98 kurz
brightness 78 Helligkeit
bring out 72 herausbringen
broadband internet connection 46 Breitband-Internetanschluss, -zugang
broadcast station 75 Sender
brochure 104 Broschüre
broken 22 kaputt, defekt
browse (through) 71 durchsuchen, (herum)stöbern (in)
brush against 93 bürsten an
BTEC (National Diploma or other technical certificate) 25 staatlicher technischer Abschluss in GB
building site 121 Baustelle
bullet point 35 Stichpunkt
business 7 Geschäft
do business with 103t Geschäfte machen mit
get down to business 105 zur Sache kommen, an die Arbeit gehen
business card 104 (Visiten-)Karte
businessman 6 Geschäftsmann
businesswoman 33 Geschäftsfrau
by means of 85 mit Hilfe von, durch

C

cabinet 64 Schaltschrank
cable 10 Kabel, Leitung
cabling 51 Verkabelung
calculate 58 kalkulieren, berechnen

calculator 77p (Taschen-)Rechner
call back 50 zurückrufen
on call 12 auf Abruf, abrufbereit
call out 89p ausrufen
make calls 95 Anrufe erledigen
cam timer 112 Nockensteuerung
cancel out 42 sich (gegenseitig) aufheben
candidate 15 Bewerber/in
canoe 124p Kanu fahren
canoeing club 124p Kanuverein
capable of 44 fähig, in der Lage zu
capacitance meter 76 Kapazitätsmesser
capacitor 66 Kondensator
captain 119 Kapitän
caption 96 Bildunterschrift
capture 53 Halte-
cardiac arrest 45 Herzstillstand
career 15 beruflich(er Werdegang), Karriere
careful(ly) 53 sorgfältig, vorsichtig
carrier tray 53 CD-Schlitten
carry 110t befördern, bei sich haben/führen
carry out 18 aus-, durchführen
cartoon 96 Karikatur
cartoonist 96 Karikaturist/in
case 8 Fall
in case of 114 im Falle von
cash on delivery 21t Barzahlung bei Lieferung, Barkauf
cause 9 verursachen, auslösen
ceiling light 30t Deckenleuchte
cell 10 Zelle
cell phone (AE = mobile phone BE) 95 Handy, Mobiltelefon
cellar 12 Keller
certain 57 gewiss, bestimmt
certainly 8 sicherlich, gewiss, bestimmt
certificate 119 Urkunde
chain reaction 39 Kettenreaktion
chair 107 Leiter/in
challenge 97 Herausforderung, Aufgabe
channel 27 Kanal, Leerrohr, Elektroinstallationsrohr
chaotic 97 chaotisch
character 105 Charakter, Figur, Person
characteristic 75 (typisches) Merkmal, Eigenschaft
charge controller 62t Ladungssteuergerät
charge 62t Ladung
in charge of 11 verantwortlich für
charred 67 verkohlt
chassis 53 Grundplatte
chat 74 Chat, Unterhaltung
check in 110 abgeben
check-in counter 110 Abfertigungsschalter
check-in desk 110 Abfertigungsschalter
checklist 102 Checkliste
cheers 71 tschüs, danke
chest 45 Brust(korb)
china 95 Porzellan
choice 8 Wahl
choir 120 Chor
choose 30 (aus)wählen
chronological order 52 zeitliche Reihen-, Abfolge
chunky 125 dick, klobig
circle 105 Kreis
circuit 15 Schaltung, (Strom-)Kreis
circuit board 70 (Schaltungs)Platine
circuit breaker 17 Stromunterbrecher, Überspannungsschalter
circuit diagram 82 Schaltbild
circulate 57 zirkulieren
city 12 (Groß-, Innen-)Stadt
claim 55 Behauptung
clamp 91 Klammer
clarification 50 Erklärung, Klärung, Erläuterung(en)
classmate 25 Klassenkamerad/in
clean up 124p sauber machen, aufräumen
clerk 110t Angestellte/r
click 15 (an)klicken
client 105 Kunde/Kundin
climate 97 Klima
closed 64 geschlossen
closure 44 (Ab-)Sperrung
clutch 88 Kupplung
coal 57 Kohle
coat 125 Mantel
coaxial 49 Koaxial-
code 55 Code
coil 42 Windung, Spule
collect 109 abholen
collector 57 Sammler/in, Kollektor
college 17 (Berufs-)Fachschule, (Fach-)Hochschule
colour 27 Farbe
colour-coded 80 farbig markiert
coloured 64 farbig
column 103 Spalte
combination 41 Kombination
combine (with) 25 verbinden, kombinieren (mit)
come from 58 stammen von
come loose 70 sich lösen
come out 72 erscheinen, herauskommen
come up 111 sich ergeben
come with 95 geliefert werden mit
command 121 Beherrschung
commercial 6 Geschäfts-, gewerblich
commissioning 107 Inbetriebnahme
commitment 25 Einsatz, Engagement
committee 97 Ausschuss
common 55 häufig, weit verbreitet
communicate 75 übermitteln
communication 13 Kommunikation(s-)
communications technology 10 Nachrichtentechnik
community service 120 Zivildienst
commutator 92 Stromwender
compact 37 kompakt
company 6 Unternehmen, Firma
compare 15 vergleichen, sich vergleichen lassen

comparison 18 Vergleich
compatibility 97 Kompatibilität
competence 96 Kompetenz
competitive 15 konkurrenzfähig
complain 95 (sich) beschweren
complete 8 vervollständigen, ergänzen
completely 27 vollständig
complex 17 kompliziert, komplex
complexity 95 Komplexität
complicated 94 kompliziert
comply with 44 befolgen
component 18 (Bau-)Element, (Bau-)Teil
comprehensive 15 umfassend
concerned about 47 besorgt wegen
conclude 90 (ab)schließen
in conclusion 113 zum Schluss, abschließend
condition 17 Bedingung
conduct away 81 ableiten, abführen
conduit 17 Kabelrohr
confident 114 sicher
configuration 50 Konfiguration, Geräteanordnung
configure 52 konfigurieren
confirm 107 bestätigen
confirmation 79 Bestätigung
conform to 103t entsprechen
connect (up) 12 (miteinander) verbinden, anschließen
connection 53 Verbindung
consensus 97 Konsens, Einigung
consequence 57 Folge, Konsequenz
consequently 71 infolgedessen
consider 15, 102 bedenken, berücksichtigen; betrachten (als)
for consideration 15 zur Beurteilung
construction 107 Bau, Konstruktion
construction site 120 Baustelle
consult 110 beachten
consultant 95 Berater/in
consumer 35 Verbraucher/in
consumer unit 26 Sicherungskasten
consumption 59 Verbrauch
contact 23 Kontakt
contactor 23 (Schalt-)Schütz
contain 58 enthalten
container 89 Behälter, Gefäß
continuity 94 (Strom-)Versorgung
continuous flow water heater 37 Durchlauferhitzer
contract 6 Vertrag
contractor 7 Unternehmer (mit Werkvertrag), Auftragnehmer/in
contradict 55 widersprechen
contrary to 103t gegenüber, im Gegensatz zu
contribution 62t Beitrag, Lieferung
control panel 17 Kontrollpult, -tafel
control 17 steuern
controller 64 Steuergerät
conventional 56 konventionell
convergence 97 Konvergenz, Zusammenpassen
converted 38 umgebaut
cooker 27 Herd
coolant flow monitor 106 Kühlmittelkontrollanzeige
cooperate 114 zusammenarbeiten
copper 10 Kupfer
cordless drill 21p Akku-Schrauber, -Bohrer
cordless phone 12 schnurloses Telefon
corner 30t Ecke
corporate executive 95 leitende/r Angestellte/r
correct(ly) 55 richtig
corresponding to 89 entsprechend
corridor 14 Gang, Flur, Korridor
at all cost 103t unter allen Umständen
cost-effective 46 kostengünstig
cotton 91 Baumwolle
counterpart 115 Gegenüber
cover letter 121 Begleitschreiben
cover (up) 15, 43 abdecken, einschließen
go crazy 84 wütend werden, ausrasten
credit card payment 21t Bezahlung mit Kreditkarte
crimp 53 Schlaufe
critical 115 kritisch
criticize 115 kritisieren
cross-cultural 103t multikulturell
crystal 95 Kristall
cup holder 95 Tassenhalter
current 23 Strom
custom 105 Sitte, Brauch
customary 103t üblich
customer 6 Kunde/Kundin
customer service 95 Kundendienst
customize 95 individuell einstellen
customs counter 110 Zollschalter
customs 102 Zoll
customs declaration card 110t Zoll(inhalts)erklärung
cut away 29 abschneiden
CV (curriculum vitae) 117 Lebenslauf
cycle 93 Zyklus, Drehung
cylinder 89 Zylinder

D

D.I.Y. (Do-it-Yourself) 12 Heimwerker(-)
DAB digital radio 10 digitaler Rundfunkempfang
damage 45 Schaden, Schäden
Danish 58 dänisch
data 46 Daten
database 104 Datenbank
day release 22 freier Tag (zur Fortbildung)
in those days 27 damals, in jener Zeit
deal with 17 sich befassen/ beschäftigen mit, bearbeiten, erledigen
decide 16 (sich) entscheiden
decimal 60 Dezimalzahl
decision making 103t Entscheidungsfindung
declare 110 verzollen, zur Verzollung anmelden

decrease 81 verringern
dedicate 72 widmen
defective 71 fehlerhaft
define 81 definieren, (genauer) erklären
definitely 57 bestimmt, sicher(lich)
definition 9 Definition
degree 59 Grad
delayed 110 verspätet
be delighted 7 sich (sehr) freuen
deliver 115 vortragen
delivery 13 (Aus-)Lieferung, Liefer-
demand (for) 25 Nachfrage (für), Bedarf (an)
dental 15 zahnärztlich
depart from 108 abfliegen, starten
department 12 Abteilung
departure 110 Abflug
depend 25 abhängen (von)
dependent 97 abhängig
depending on 17 je nachdem
deplete 57 erschöpfen
deploy 55 einsetzen
depress 64 (nieder)drücken
descending 115 absteigend
description 12 Beschreibung, Schilderung
be designed to 23 zu etw dienen
designer 26 Konstrukteur/in, Designer/in
desired 85 gewünscht
desoldering pump 71 Entlöt(saug)pumpe
despite 97 trotz
destroy 115 zerstören
detach 91 abnehmen
detail 7 Einzelheit, Detail
detailed 104 detailliert, ausführlich
determine 61 bestimmen, feststellen
develop 22 entwickeln, bekommen
developing 97 Entwicklungs-
development 88 Entwicklung
device 18 Gerät, Apparat
dew 64 Tau
diagnose 17 diagnostizieren
diagram 13 grafische Darstellung, Grafik
dial 55 wählen
diameter 65 Durchmesser
differ 101 sich unterscheiden
digital camera 95 Digitalkamera
digital display 95 Digitalanzeige
diode 66 Diode
give directions 14 den Weg beschreiben
directive 97 Richtlinie, Anordnung
directly 25 direkt
director 11 Direktor/in, Leiter/in
dirty 65 schmutzig
disadvantage 15 Nachteil
disagree 19 anderer Meinung sein, widersprechen, nicht zustimmen
disaster 27 Katastrophe
discoloured 71 verfärbt
disconnect 42 unterbrechen, trennen
discover 71 entdecken, feststellen
discuss 7 besprechen, erörtern
discussion 15 Diskussion, Gespräch, Erörterung
disengaged 88 nicht betätigt
dishwasher 30t Geschirrspüler
disk brake 89 Scheibenbremse
disparity 100 Ungleichheit
display 78 (an)zeigen
disposable 65 Einweg-, Wegwerf-
distance 9 Entfernung, Strecke
distortion 75 Verzerrung
distributed 103t verteilt
distribution 6 Vertrieb, (Aus-)Lieferung
disturb 45 stören
divide 61 teilen, dividieren
division 78 Abteilung
DMM (digital multimeter) 21p digitales Vielfachmessinstrument
domestic 57 häuslich, Haushalts-
domestic appliance technician 89p Hausgerätetechniker/in
domestic household 16 Privathaushalt
doorbell 26 Türklingel
doorway 97 (Haus-)Eingang, Tür
double 30t doppelt, Doppel-
double pole 31 zweipolig, Zweipol-
double-check 79 (noch einmal) nachprüfen
doubt 19 zweifeln
download 92 herunterladen
drain 88 entleeren
dramatic 95 dramatisch
draw up 12 abfassen, aufstellen
drawing 17 Zeichnung
dressing table 31 Frisierkommode
drill 21 bohren
driver's license (AE = driving licence BE)
driving licence 108 Führerschein
drop 82 Rückgang
dry cleaners 65 chemische Reinigung
durability 57 Dauer(haftigkeit)
duration 45 Dauer
DVB digital TV 10 digitaler Fernsehempfang

E

earn 116 verdienen
earth 27 Erde; erden
earth conductor 27 Schutzleiter
earth terminal 26 Erdpol, Masseklemme
earth wire 27 Schutzleiter
earthed socket outlet 31 geerdete Steckdose, Schukosteckdose
eco-friendly 65 umweltfreundlich
economic 75 wirtschaftlich
economical 88 sparsam, wirtschaftlich
effective(ly) 103t wirkungsvoll
effectiveness 91 Effektivität
efficiency 58 Effizienz

efficient(ly) 37 effizient, (leistungs)fähig
elderly 32 älter
electric 11 elektrisch, Elektro-
electric cooker 27 Elektroherd
electric drill 21p Elektrobohrer
electric shock 40 elektrischer Schlag
electrical 7 elektrisch, Strom-
electrical circuit 21 Stromkreis
electrical contractor 26 Elektroinstallateur
electrical engineering 124p Elektrotechnik
electrical fitter 117 Elektroinstallateur/in
electrical plant 122 Elektroanlage
repair and maintenance electrician 16 Elektrotechniker/in für Reparatur und Wartung
electrician 24 Elektriker/in, Elektrotechniker/in
electrician's mate 120 Elektrotechnikgehilfe, -in
electricity meter 26 Arbeitszähler
electricity pylon 35 Strommast
electrocute oneself 34 sich durch Stromschlag töten
electrolytic capacitor 80 Elektrolytkondensator
electromagnet 93 Elektromagnet
electron 57 Elektron
electronic 15 elektronisch
electronic servicing 25 elektronische Wartung
electronic(s) technician 6 Elektroniker/in
eliminate 82 eliminieren, beseitigen
else 30t sonst (noch)
emergency 17 Notfall
emission 63 Emission
emit 65 emittieren, ausstoßen
employ 11 beschäftigen
employee 6 Angestellte/r
employer 25 Arbeitgeber/in, Unternehmer/in
employment 15 Arbeit, Beschäftigung
en route 110 unterwegs
enable 59 (es) ermöglichen
Enc (enclosed) 121 Anlage
enclose 121 beifügen
enclosure 64 Umhüllung, Ummantelung
at the end of 14 am Ende des/der
energize 89 aufladen
energy 10 Energie
energy efficient 119 energiesparend
engaged 88 eingelegt
engine 97 Motor
engineer 12 Ingenieur/in, Techniker/in
engineering 15 Technik, Maschinenbau
English-speaking 14 englischsprachig
enjoy 17 gern haben/tun
enquiry 90 Anfrage
enthusiastic(ally) 115 begeistert, enthusiastisch
entirely 62t ganz, vollständig
environment 56 Umwelt
environmental 57 Umwelt-
environmentally friendly 59 umweltfreundlich
equal 19 gleich (groß)
equipment 12 Ausrüstung, Geräte, Einrichtung
equivalent 18 Entsprechung
error 8 Fehler, Irrtum, Versehen
escape (from) 30, 89 ausfließen aus, entweichen
especially 17 besonders
essential 103t (absolut) notwendig
establish 55 herstellen
estimate 37 Kostenvoranschlag
evening wear 125 Abendgarderobe
event 107 Ereignis, Vorgang
ever 19 je(mals), schon (ein)mal
exact 25 genau
examine 74 untersuchen, prüfen
exceed 93 übersteigen, -schreiten
excellent 7 ausgezeichnet, hervorragend
except 64 außer
exceptional(ly) 85 außergewöhnlich
in excess of 85 über (hinaus)
excessive 53 übermäßig
exchange student 79 Austauschstudent/in
exciter field 93 Erregerfeld
exclusive(ly) 44 ausschließlich, alleinig
executive 103t leitende/r Angestellte/r, Manager/in
exhaust 89 Auspuff
exhibitor 104 Aussteller/in
expand 41 sich ausdehnen
expect 9 erwarten, schätzen, annehmen
experience 7 (Berufs-)Erfahrung
experiment 82 Versuch, Experiment
expertise 88 Fach-, Sachkenntnis
expiry date 122 Ablaufdatum
explanation 64p Erklärung
exploit 57 be-, ausnutzen
explosion-proof 8 explosionssicher
explosive 110t Sprengstoff
exposed 64 freiliegend
expression 14 Ausdruck
extend 64 verlängern, ausziehen
extension 37 Verlängerung
extensive 104 umfassend, umfangreich
external 45 äußer(lich)
extractor fan 30t Abzugshaube
extra-curricular activity 118 außerschulische Aktivität
extreme(ly) 45 äußerst, außerordentlich

F

fabricate 15 herstellen, fertigen
facility 53 Vor-, Einrichtung
facing 57 in Richtung
in fact 17 tatsächlich, eigentlich
factor 59 Faktor
factory 12 Fabrik
failure 105 Fehlschlag, Versagen
fair 104 Messe
fairly 48 ziemlich, recht
fall into 34 hineinfallen
fall onto 34 fallen auf
familiar with 119 vertraut mit
familiarize with 52 (sich) vertraut machen mit

famous 67 berühmt
FAQs 63 häufig gestellte Fragen
faulty 23 fehlerhaft, defekt
in favour of 19 für (etwas)
feature 64 Merkmal
federal 57 Bundes-
Federal Agency for Technical Relief 123 Technisches Hilfswerk
feed (into) 35 eingeben (in), einspeisen (in)
feeder 106 Zubringer, Einzug
feel free 103 sich nicht scheuen
fence off 43 abschranken
fetch 124p holen, bringen
fibre optic 10 Glasfaser
figure 40 Abbildung; Zahl, Ziffer
file 21 Akte, (Akten-)Ordner
filter out 95 herausfiltern
final 33 letzte/r/s
final grade 124 Abschlussnote
finite 57 begrenzt, endlich
fire 12 feuern, entlassen
firearm 110t Schusswaffe
firework display 101 Feuerwerk
first aid 45 erste Hilfe
fit 44 installieren, einbauen
fittings 8, 30t Anschlüsse, Armaturen
five-core cable 30t fünfpolige Leitung
fix 17 befestigen
fixed 17 fest(gesetzt)
fixture 38 (Elektro-)Installation
flat 40 Wohnung
flat-bladed screwdriver 20 Schlitzschraubenzieher
flex 26 Netzkabel
flexible 46 flexibel
floating operation 85 erdfreier Betrieb
flow along 40 durchfließen, -strömen
flow meter 111 Durchflussregler
flow through 40 durchströmen, -fließen
fluent 118 fließend
fluid 57 Flüssigkeit
fly with 108 fliegen mit
focus 78 scharf einstellen, bündeln
folding rule 20 Zollstock
following 6 folgende
foodstuff 110t Nahrungsmittel
force 57 zwingen
forceful(ly) 115 kräftig
forever 58 ewig
forward 71 vorwärts
fossil fuel 57 fossiler Brennstoff
found 97 gründen
fourfold 31 vierfach
fraction 60 Bruch(teil)
freestanding 91 freistehend
freeze 45 erstarren
freezing 45 erstarrend
frequency 75 Frequenz
frequent 88 häufig, oft
fridge-freezer combination 30t Gefrierkombination
front door 64 Vordertür
in front of 14 vor
front panel indicators 64 Anzeigen der Frontplatte
become frozen 44 einfrieren
frustrate 55 frustrieren
full-wave bridge rectifier 66 Zweiwegbrückengleichrichter
make fun of 115 sich lustig machen über
function 17 funktionieren, arbeiten
function test 23 Funktionsprüfung
further 7 weitere
fuse 17 Sicherung
fuse box 27 Sicherungskasten
integrated voltage regulator 83 integrierter Spannungsregler
fused 44 abgesichert
future 57 Zukunft

G

gain 55 gewinnen, erlangen
gap 9 Lücke
garlic 125 Knoblauch
gas cooker 27 Gasherd
gate 110 Flugsteig, Ausgang
gauge 97 Spurweite
in forward gear 88 im Vorwärtsgang
in reverse gear 88 im Rückwärtsgang
gender 119 Geschlecht
general 36 allgemein
generalize 103t verallgemeinern
generate 35 erzeugen
generator 12 Generator
generous 117 großzügig
gentle 103t sanft, leicht
geothermal 56 Erdwärme-
German higher education entrance qualification 120 Fachhochschulreife
German university entrance qualification 120 Hochschulreife (Abitur)
get the wrong end of the stick 43 etw falsch verstehen
get-it-done 103t ergebnisorientiert
GFCI (ground fault circuit interrupter) (AE = RCD) 70t Fehlerstromschutzschalter
give an electric shock 34 einen Stromschlag verpassen
give in 125 abgeben
be glad 114 sich freuen
at a glance 91 auf einen Blick
global warming 63 Erderwärmung, Erwärmung der Erdatmosphäre
go back and forth 65 hin und her gehen
go off 110 losgehen, ansprechen
goods 9 Ware(n), Güter
goods vehicle licence 120 Lkw-Führerschein

government 63 Regierung
government-sponsored 15 staatlich gefördert
gradation 91 Einteilung
grade 123 Note
graph 78 Kurve, Diagramm
grateful 121 dankbar
greenhouse effect 57 Treibhauseffekt
greeting 103t Begrüßung
grid 57 Stromnetz
grid-tied 62 an das Stromnetz angeschlossen
grinding machine 22 Schleifmaschine
grip 21 (er)greifen
ground 40 Erdung
grounding contact 101 Erdkontakt
grounding pin 101 Erdstift
guarantee 25 garantieren
guess 12 (er)raten
guide 44 Richtlinie, Einführung
guided tour 13 Besichtigung, Führung
guideline 95 Richtlinie, Hinweis
guy 72 Kerl, Typ

H

hacksaw 20 Bügelsäge
hair dryer 93 Föhn
ham radio operator 67 Amateurfunker/in
hammer 20 Hammer
hand luggage 110 Handgepäck
handheld organizer 95 Taschencomputer
handout 102 Merkblatt
handover 111 Übergabe
handshaking 103t Händeschütteln
happen 8 passieren, geschehen
hard minerals 95 feste Mineralien
hardware 46 Geräte
hardwired 112 fest verdrahtet
harm 40 Schaden
harmonize 97 harmonisieren, abstimmen
harness 91 Verkleidung
hasty 55 hastig, überhastet
HD television 97 HD-Fernsehen
head office 107 Zentrale
heart 45 Herz
heartbeat 45 Herzschlag
heat 40 Hitze, Wärme
heat up 57 erhitzen, erwärmen
heater coil 89 Heizspirale
heating 17 Heizung
hedge 38 Hecke
height 17 (große) Höhe
held up by 35 gehalten durch
helpline 25 Telefonberatung, telefonische Hilfe
hierarchy 103t Hierarchie
high access platform 17 Arbeits-, Hebebühne
higher education entrance qualification 120 Fachhochschulreife
highest-ranking 103t höchstrangig
high-power RF application 75 Anwendung, Einsatz im Hochleistungs-Radiofrequenzbereich
high-speed 93 Hochgeschwindigkeits-
hire 12 ein-, anstellen
hold 91 (fest)halten
hole 21 Loch
horizontal (X) axis 78 horizontale (X-)Achse
host 115 Gastgeber/in
hotline 10 Hotline, Kundentelefon
hotplate 65 Koch-, Warmhalteplatte
household 11 Haushalt(s-)
housing 57 Gehäuse, Verkleidung
however 41 jedoch
hub 52 Verteiler(kasten)
human resources 11 Personal(abteilung)
hybrid 75 hybrid
hydraulic 112 hydraulisch
hydro-electric 56 Wasserkraft-
hydrogen 63 Wasserstoff

I

ideal 15 ideal
identical 42 identisch
identify 66 identifizieren, festlegen
IEC (International Electrotechnical Commission) 96 Internationale Elektrotechnische Kommission
ignorance 105 Unwissenheit, Unkenntnis
ill 34 krank
illuminate 67 beleuchten
illustrate 103t illustrieren, veranschaulichen
immediately 12 sofort, umgehend
immersion heater 37 Warmwasserspeicher
impact 55 (Aus-)Wirkung
implement 55 einsetzen, installieren
impolite 115 unhöflich
importance 98 Wichtigkeit, Bedeutung
impossible 43 unmöglich
impractical 55 unpraktisch, unbrauchbar
impression 50 Eindruck
impressive 111 beeindruckend, eindrucksvoll
improve 97 verbessern
incentive 57 Anreiz
incident 55 Vorfall, Ereignis
inclination 59 Neigung
include 15 umfassen, einschließen
including 12 einschließlich, inbegriffen, darunter
incomplete 53 unvollständig
inconvenient 17 unbequem
incorporate 64 umfassen
increase 40 Zunahme, Steigerung
increasing(ly) 97 zunehmend
independent 62t unabhängig
index finger 105 Zeigefinger
indicate 89 (an)zeigen
indoors 18 drinnen, in geschlossenen Räumen
industrial 6 Industrie-, industriell

industrial plant 16 Industriewerk, -anlage
industrialized 97 Industrie-
inefficiency 61 Unzulänglichkeit
inefficient 46 unwirtschaftlich
inflexible 98 unflexibel
inform 25 informieren, unterrichten
informal 32 formlos, locker
information desk 110 Informationsschalter, Auskunft
infrared 57 infrarot
infrastructure 55 Infrastruktur
initially 112 anfänglich, zuerst
initiate 64 einleiten
initiative 63 Initiative
ink cartridge 95 Tintenpatrone
innumerable 103t unzählig, zahllos
input 67 Eingang (technisch)
inquire about 104 sich erkundigen nach
insecurity 125 Unsicherheit
insist on 75 bestehen auf
insolation 61 Sonnenbestrahlung
inspect 16 kontrollieren, überprüfen
install 8 installieren, einbauen
installation 7 Installation, Einbau
installation electrician 16 Elektroinstallateur/in
installer 44 Installateur/in
instant(ly) 115 sofort, umgehend
institute 103t Institut
instruct 111 einweisen, instruieren
instruction manual 95 Bedienungsanleitung, Benutzerhandbuch
instruction 18 Anweisung, (Betriebs-)Anleitung
insulated 57 isoliert
insulation 29 Isolierung, Isolation
insult 115 beleidigen
integrated 80 integriert
integrated voltage stabilizer 66 integrierter Spannungsstabilisierer
be intended for 85 dienen zu
intensity 78 Intensität
interact with 93 sich gegenseitig beeinflussen
interaction 93 Interaktion
interconnect 59 miteinander verbunden sein
be interested in 15 interessiert sein an, sich interessieren für
interface 112 Schnittstelle
interfere 98 sich einmischen, stören
intermediate switch 31 Zwischenschalter
internal 45 inner(lich)
internet connection 46 Internetanschluss, -zugang
interrupt 103 unterbrechen
interview 15 Vorstellungsgespräch
interviewer 123 Leiter/in des Vorstellungsgespräches, Interviewer/in
introduce (oneself) 8, 32 (sich) vorstellen, miteinander bekannt machen
invent 12 erfinden
inverter 23 Wechselrichter
invite 15 einladen
invoice 21t eine Rechnung ausstellen (auf), berechnen
involve 16 einschließen, umfassen
ISO (International Organization for Standardization) 97 Internationale Organisation für Normung
issue 67 Ausgabe
item 17 Gegenstand, Ding
ITU (International Telecommunications Union) 97 Internationale Fernmeldeunion

J

jewellery 125 Schmuck
job profile 18 Stellenbeschreibung, Berufsbild
join 57 sich anschließen, teilnehmen an
joke 95 Witze machen
junction box 26 Verteilerkasten
jungle 97 Dschungel
junior 114 untergeordnet
in mid-journey 97 mitten auf der Reise

K

key point 103 Kernpunkt, -frage
Key Skills qualification 25 Bescheinigung über Schlüsselqualifikationen
keyword 103 Stichwort
kill oneself 34 sich töten
Kind regards 52 Mit freundlichen Grüßen
kitchenette 37 kleine Küche, Kochnische
knob 78 Knopf
knock loose 57 freisetzen
know-how 10 Fachkenntnis, Sachverstand
known as 30 bekannt als

L

label 62 beschriften
ladder 34 Leiter
LAN (local area network) 12 lokales Netz
at the latest 107 spätestens
launch 104 einführen, herausbringen
laundry 65 Wäsche, Wasch-
layout plan 47 Grundriss
leading 71 führend
leaflet 104 Prospekt
leak 89 undichte Stelle, Leck
leakage 40 Austreten, Ausströmen
at the very least 125 zumindest
leave 22 abgehen (von), verlassen
leave for 108 abreisen nach
leaving certificate 119 Abschlusszeugnis
LED (light emitting diode) indicator 64 LED (Leuchtdiode)n-Anzeige
on your left 14 zur Linken
length 29 Länge

let go of 45 loslassen
let sb know 109 jdm Bescheid geben/sagen, jdn informieren
licensed 117 lizensiert
lid 95 Deckel
life-saving 119 Lebensrettungs-
light switch 26 Lichtschalter
lighting 8 Beleuchtung, Licht
lighting element 8 Leuchtelement
lighting outlet 31 Lichtanschluss
lighting technology 10 Lichttechnik
like poles 93 gleiche Pole
likely 45 wahrscheinlich
limit (to) 100 begrenzen, beschränken (auf)
limiting 85 Begrenzung
in line 101 in (einer) Reihe
link 13 Verbindung, Link
liquid 40 Flüssigkeit
live 43 unter Spannung/Strom (stehend)
load 23 Last, Belastung
local 12 örtlich, lokal
locate 17 ausfindig machen
located 6 gelegen, mit Sitz in
location 15 (Stand-)Ort
lock 29 schließen
locomotive 97 Lok(omotive)
look forward to 7 sich freuen auf
take a look at 24 (sich) ansehen
look up 71 heraussuchen
lose face 115 das Gesicht verlieren
lost 42 verloren (gegangen)
luckily 67 glücklicherweise, zum Glück
luggage 110 Gepäck
lyrics 72 (Lied-)Text

M

machinery 106 Maschinen(park), Geräte
madam 32 gnädige Frau
be made of 41 bestehen aus
mains input socket 66 Stromverteilerdose, Netzanschlussdose
mains tester 76 Spannungsprüfer
the mains 43 Stromnetz
maintain 16 warten, pflegen
maintenance 13 Wartung
maintenance electrician 15 Wartungstechniker/in
maintenance personnel 122 Wartungspersonal
make up 75 ausmachen
malfunction 17 nicht funktionieren
malicious 55 böswillig, mutwillig
manage 49 umgehen (mit)
managing director 7 Geschäftsführer/in
manual 89p Handbuch
marital status 119 Familienstand
mark 30t markieren, kennzeichnen
marketing 95 Werbe-, Vertriebs-, Marketing(-)
massive 40 riesig, enorm
master electrician 6 Elektro(technik)meister/in
meaning 48 Bedeutung
meanwhile 57 inzwischen
measure 21 (ver)messen
measure up 65 den Erwartungen entsprechen
measurement 76 Maß
mediate 114 vermitteln, dolmetschen
medical 15 medizinisch
meditation 65 Meditation
melt 39 schmelzen
member 115 Mitglied
memory 112 Speicher
mention 27 erwähnen
message 17 Mitteilung, Hinweis
take a message 95 eine Nachricht aufnehmen
metal detector 110 Metallsuchgerät
meteorological 59 meteorologisch
meter 62t Zähler
methane 56 Methan
method 61 Methode, Verfahren
microwave 27 Mikrowelle
in the middle 31 in der Mitte
mild 45 mild, leicht
military service 120 Militärdienst
millisecond 41 Millisekunde
mind 17 etw dagegen haben
minutes 106 Protokoll
misconception 55 falsche Annahme
misinterpretation 103t Missverständnis
missing 27 fehlend, nicht vorhanden
mix up 10 verwechseln, durcheinander bringen
mobile 9 Handy, Mobilfunk
mode 64 Modus
modem 46 Modem
moderate 45 mäßig
modern 10 modern
modernize 10 modernisieren
module 53 Modul
moisture 64 Feuchtigkeit
monitoring 107 Überwachung, Kontrolle
monolithic 85 monolithisch
monotonous 12 monoton
mostly 17 hauptsächlich, meistens
motor 11 Motor
motor coupling 91 Motorkupplung
motor shaft 91 Motorwelle
motor starter 89 Startermotor
motor vehicle 98 Kraftfahrzeug
motorway 17 Autobahn
mount bracket 53 Montageträger
mounted 72 montiert
mounting 29 Montage
mounting clamp 91 Montageklammer
move on to 103 übergehen zu
mow the lawn 38 den Rasen mähen
multimedia 97 Multimedia-
multimeter 20 Vielfachmessinstrument
multiply 61 multiplizieren

muscle 45 Muskel

N

nail 125 (Finger-)Nagel
nappy 65 Windel
nasty 34 schlimm
national grid 35 Energieversorgungsnetz
natural gas 57 Erdgas
nearby 43 benachbart, in der Nähe
necessary 30 nötig, erforderlich
neck 125 Hals
necklace 125 (Hals-)Kette
needle nose pliers 20 Spitzzange
negotiable 117 Verhandlungssache
neighbour 43 Nachbar/in
neither 88 auch nicht, keine
nerve 45 Nerv
network 13, 46 Netz(werk), Sendenetz; vernetzen
network card 52 Netzwerkkarte
network technology 10 Netzwerktechnik
neutral 32 neutral
neutral coil 42 Spule im Neutralleiter
never mind 110t macht nichts
next door to 14 neben
next to 31 neben
nod 103t nicken
noise 57 Geräusch, Lärm
nominal current 41 Nennstrom
nominal value 40 Nennwert, Nominalwert
non-polluting 57 ohne Schadstoffe
non-productive 103t unproduktiv
make a note of 43 notieren, beachten
make notes 24 (sich) Notizen machen
at short notice 121 kurzfristig
nowadays 14 heutzutage
nowhere 27 kein Platz/Ort
nuclear 56 Atom-
nut 21 (Schrauben-)Mutter
NVQ (National Vocational Qualification) 22 staatlicher Berufsabschluss in GB

O

objective 107 Ziel, Zweck
observe 103t beobachten, beachten
obvious 71 offensichtlich, sichtbar
occur 17 vorkommen, eintreten
offend 105 beleidigen
offline 122 offline, nicht übers Internet
offset 65, 101 ausgleichen; nicht in einer Reihe
offshore 122 vor der Küste
ohmmeter 76 Ohmmeter, Widerstandsmesser
oil refinery 18 Ölraffinerie
oil 57 Öl
online 15 online, im/übers Internet
operate 17 arbeiten, laufen, funktionieren
operating instruction 64 (Bedienungs-)Anleitung
operating manual 64 Handbuch, Bedienungsanleitung
operating system 52 Betriebssystem
operating theatre 121 Operationssaal
operator 24 Maschinist/in
opportunity 15 Chance, Möglichkeit
opposing 93 entgegengesetzt
optical 15 optisch, augenärztlich
option 49 Option, (Wahl-)Möglichkeit
optional 15 wahlweise
order 21 bestellen
ordinary 98 gewöhnlich, normal
organizer 104 Organisator/in, Veranstalter/in
organization chart 11 Organisationsübersicht, -struktur
orientation 59 Orientierung, Richtung
oscillation 75 Schwingung
oscilloscope 76 Oszilloskop
out of adjustment 89 falsch eingestellt
outage 57 Ausfall
outlet 31 Anschluss
outline 25 Überblick, Abriss
output 67 Ausgang (technisch)
outstanding 57 hervorragend
over weight 110 Übergewicht
overcharge 62t überladen
overhead 34 an der Decke
overland 35 über Land
overload 41 Überlastung
overloaded 23 überlastet
overseeing 97 Aufsichts-
overtime 15 Überstunden
owner 63 Besitzer/in, Eigentümer/in

P

pack 110t (ein)packen
package 25 Bündel
packing 8 Verpackung, Pack-
pain 45 Schmerz(en)
painful 45 schmerzhaft
paint 6 Farbe, Lack
a pair of 21 ein Paar, eine (Zange)
pairwork 21 Partnerarbeit
PAN (Personal Area Network) 97 Kleinstnetzwerk zur persönlichen Kommunikation
pan 65 Topf, Pfanne
panel 15 Anzeige-, Schalttafel
panel building electrician 16 Schalttafel(bau)elektriker/in
pants (AE = trousers BE) 95 Hose(n)
paper clip 70 Büroklammer
paragraph 39 Absatz, Abschnitt
parallel 101 parallel
part of 29 Teil von
participant 25 Teilnehmer/in

particular 55 besonder(s), speziell
pass through 30 durchlaufen, passieren
pass 22 (Examen) bestehen, ablegen
passenger 110 Passagier/in, Fluggast
patchpanel 52 Patchpanel, Verteilerfeld
patient 94 geduldig
pattern 103t Muster
payment terms 21p Zahlungsbedingungen
peak voltage 77 Spitzenspannung
peak 61 Gipfel, Spitze, Höchstwert
peer 115 Kollege/-in
pension 15 Rente, Pension
percentage 55 Prozentsatz
perfect(ly) 95 vollkommen
perform 7 aus-, durchführen
performance 91 Leistung(en)
periodical(ly) 17 regelmäßig
peripheral 46 Peripherie(gerät)
permanent 36 ständig
in person 7 persönlich
personal comfort zone 103t Diskretionsabstand
personal contact 103t körperliche Berührung
personal details 119 persönliche Angaben
personnel 64 Personal, Mitarbeiter
persuasive 55 überzeugend
petrochemical 122 petrochemisch, Erdöl-
petrol station 8 Tankstelle
phase 40 Phase
phase coil 42 Spule im Phasenleiter
Phillips screwdriver 20 Kreuzschlitzschraubendreher
phone card 84 Telefonkarte
phrase 9 (Rede-)Wendung, Ausdruck
physical 42 physikalisch
pick up 21 aufheben, aufnehmen
pick 18 auswählen
pin 100 Stift, Nadel
pipe 27 Rohr(leitung)
pitfall 105 Falle, Fallstrick
place an order 21t einen Auftrag erteilen, eine Bestellung aufgeben, bestellen
placement machine 92 Bestückungsmaschine
planet 63 Planet
plant 65 pflanzen
plastic 29 Plastik, Kunststoff
plate 57 Platte
PLC (programmable logic controller) 17 speicherprogrammierbare Steuerung
plenty of 117 viel, reichlich
plug 8 Stecker
plug in 71 (Stecker) einstecken
pneumatic 112 pneumatisch
point of view 19 Standpunkt, Ansicht
point out 19 hinweisen auf, aufmerksam machen auf
pointed 100 spitz
policy 111 (Firmen-)Politik
polite 32 höflich
politeness 50 Höflichkeit
pollution 57 Verschmutzung
polystyrene cup 65 Plastiktasse
popular 14 beliebt, populär
popular(ly) 101 allgemein
portion 57 Portion, Anteil
possible 45 möglich
post 71 verschicken
potentiometer 80 Potentiometer
power (a machine) 30 (ein Gerät) antreiben
power cable 34 Stromkabel, -leitung
power cord 53 Netz-, Stromkabel
power drill 20 elektrische Bohrmaschine, Elektrobohrer
power generation 25 Stromerzeugung
power output 23 Strommenge, -stärke
power rating 94 Stromverbrauch
power station 18 Kraftwerk
power supply 71 Stromanschluss
power tube 75 (Sende)röhre, Röhre mit hoher Leistung
power wire 94 (flexibles) Stromkabel
powered by 75 betrieben mit
practical 22 praktisch
predate 97 zeitlich vorangehen
prefer 56 bevorzugen
preferably 15 am besten, vorzugsweise
pre-installation 114 Vorinstallation
premises 120 Geschäft(sräume)
preparation 107 Vorbereitung
prepare (to) 19 (sich) vorbereiten (auf)
at present 58 momentan, gegenwärtig
presentation 12 Vorführung
preserve 57 erhalten
press 23 drücken
pressure 44 Druck
prevent 55 (ver)hindern, verhüten
previous 83 vorhergehend, vorherig
price tag 125 Preisschild
primary education 119 Grundschul(aus)bildung
principle 57 Grundsatz, Prinzip
printer 46 Drucker
private 6 privat
probe 94 Sonde
problem solving 25 Problemlösung
procedure 91 Verfahren
proceed to 110 sich begeben nach
process 55 Prozess, Ablauf, Vorgang
processor 53 Prozessor
producer 95 Produzent/in, Hersteller/in
production line 18 Fließ-, Fertigungsband
productive(ly) 97 produktiv
programmable 112 programmierbar
programmer 99 Programmierer/in
progress 107 Fortschritt/e
project manager 107 Projektmanager/in
prong 91 Zinke, Zacke, Spitze, Dorn
proper 53 ordnungsgemäß, angemessen
properly 17 richtig, korrekt
property 97 Eigentum
proprietary 112 Eigentümer-, Besitzer-

protect 19 (be)schützen
protection 55 Schutz
protective 27 Schutz-
protective earth 30 Schutzleiter
provide 44 liefern
provider 62t Anbieter, Versorger
public 57 öffentlich
publish 97 veröffentlichen, herausgeben
pull 34 ziehen (an)
pulsate 82 pulsieren
pump 91 Pumpe
punctual 123 pünktlich
purchase, purchasing 11 (Ein-)Kauf, Einkaufs-
purchaser 95 Käufer/in
purpose 110 Zweck
push button 31 Drucktaste
put in touch 10 verbinden
PV (photovoltaic) cell 56 Photovoltaik-Zelle
PV (photovoltaic) system 10 Photovoltaik-Anlage

Q

qualification 25 Qualifikation, (Abschluss einer) Ausbildung
qualified electrician 6 qualifizierte/r Elektriker/in, Elektrotechniker/in
quote 122 angeben, nennen

R

rack 53 Gestell, Ständer
radiation 57 Strahlung
radiator 57 Heizkörper
radio enthusiast 67 begeisterer Amateurfunker
radio shack 38 Funkerbude
radio station 75 Rundfunkstation, Radiosender
rainfall 59 Niederschlag, Regen
raise 103t wecken
range 75 Bereich, Spektrum
range from … to 121 sich von … bis erstrecken
rank 102 einstufen, anordnen
rarely 58 selten
rate 15 Rate, (Stunden-)Satz
rated 44 Nenn-
rather than 65 anstatt
RCD (Residual Current Device) 26 Fehlerstromschutzschalter
reach 89 erreichen
react (to) 40 reagieren (auf)
read out 76p (laut) vorlesen
read through 95 durchlesen
reading 23 Anzeige
realistic 46 realistisch
realize 67 sich klar werden/sein
reasonable 54 angemessen, günstig
receive 6 erhalten, bekommen
recent 57 jüngst
reception 13 Empfang
receptionist 48 Empfang(smitarbeiter/in)
recess 29 Vertiefung
recharge 79t aufladen
rechargeable 79t aufladbar
recognize 122 anerkennen
recommend 36 empfehlen
reconnect 71 wieder anschließen
record 74 (Schall-)Platte
recording studio 74 Aufnahmestudio
recruitment agency 117 Arbeits-, Stellenvermittlung
rectangular 101 rechteckig
rectifier 71 Gleichrichter
rectify 78 gleichrichten
redesign 65 neu entwerfen, umstrukturieren
reduce 8 verringern, senken
reduction 65 Verringerung, Senkung
Ref. (reference) 121 Akten-, Bezugszeichen, Betreffzeile
refer to 30, 59 sich beziehen auf, vergleichen mit
referee 119 Referenz
reference 117 Akten-, Bezugszeichen
reflect 57 reflektieren
refrigeration 106 Kühlung
refuse 95 (es) ablehnen, sich weigern
register 117 registrieren, eintragen lassen
regulation 8 Verordnung, Vorschrift
regulator 79t Regler
reject 50 ablehnen
relationship 103t Beziehung, Verbindung
relay 112 Relais
release 74 herausbringen, veröffentlichen
relevant 58 relevant, wichtig
reliability 57 Zuverlässigkeit
reliable 10 zuverlässig
relief 123t Hilfe
rely on 58, 103t angewiesen sein auf, sich stützen auf
remain 64 bleiben
remind 107 (jdn an etw) erinnern
remove 67 entfernen, abnehmen
renewable 58 erneuerbar
rep. (representative) 109 Vertreter/in
repair 11 reparieren
repel 92 (sich) abstoßen
replace with 17 ersetzen durch, austauschen gegen
re-polish 125 nochmals putzen
report 9, 107 Bericht; berichten
represent 91 darstellen
representative 104 Vertreter/in, Repräsentant/in
reputation 7 Ruf
request 50 Bitte, Anfrage, Aufforderung
require 15 suchen
required 15 erforderlich
requirement 47 Erfordernis, Anforderung
rescue 119 Rettung(s-)
reservoir 57 Reservoir
reset 64 zurückstellen, zurücksetzen (technisch)
residual 42 Rest-, restlich
resistance 40 Widerstand
resistive divider 85 Widerstandsspannungsteiler

resistor 66 Widerstand
resource 55 Hilfsmittel, Ressource
respect 103t (be)achten, respektieren
respond 115 antworten
responsibility 55 Verantwortlichkeit
responsible for 17 verantwortlich für, zuständig für
restart 23 neu starten
restrict 55 be-, einschränken
result from 45 resultieren aus, sich ergeben aus
résumé 15 (tabellarischer) Lebenslauf
retailer 95 Einzelhändler/in
retreat 65 Rückzug, Zufluchtsort
reversal 93 Umkehr(ung)
reverse 71 Rück-, entgegengesetzt; rückwärts
review 107 Besprechung, Überblick
revision 112 Änderung, Überarbeitung
reward 25 Belohnung, Lohn
rewarding 117 lohnend, befriedigend
rewire 10 neu verkabeln
re-write 104 umschreiben, neu schreiben
rhythm 45 Rhythmus
rigid 97 streng, fest
ripple 82 Welligkeit (technisch)
rise 40 (an)steigen
risk 64 Risiko
RMS (root mean square) voltage 77 Effektivspannung
robot 123 Roboter
robotic 106 Roboter-
role-play 32 als Rollenspiel durchführen
roller bearing 23 Kugellager
roof 59 Dach
rotating speed 23 Drehgeschwindigkeit
rotation 93 Drehung
rotor 92 Rotor, Läufer
round nose pliers 20 Rundzange
round up to 61 aufrunden
routine 18 Routine(arbeit), routinemäßig
RPM (revolutions per minute) 111 (Um-)Drehungen pro Minute
rubber union 91 Gummiverbindung
rule 36 Vorschrift, Regel
run out 57 zu Ende gehen
running 44 laufend
rush 30t sich beeilen

S

salary 116 Gehalt, Lohn
sales department 47 Verkauf(sabteilung)
sales division 9 Verkaufsabteilung
salesperson 21p Verkäufer/in
satellite TV 27 Satellitenfernsehen
satisfaction 7 Zufriedenheit
satisfactory 44 zufrieden stellend, befriedigend
satisfied with 111 zufrieden mit
save 34 retten
scale 91 Skala
on schedule 7 im Zeitplan, planmäßig
schematic 62 schematisch(e Darstellung), Grafik
scheme 25 Programm, Projekt
school leaving certificate 120 Mittlere Reife
scientific 93 wissenschaftlich
screen 78 Bildschirm
screw 21 Schraube
screw up 74 zusammenschrauben
screwdriver 20 Schraubendreher, Schraubenzieher
take a seat 123t Platz nehmen
seated 115 platziert
secondary education 119 höhere Schul(aus)bildung
section 11 Abteilung, Abschnitt
secure 46 sicher
security 51 Sicherheit
security control 110 Sicherheitskontrolle
security gate 23 Sicherheitstor, -ventil
seek 119 suchen
seeming(ly) 101 anscheinend
select 29 auswählen
sell back to 63 verkaufen an
semiconductor 57 Halbleiter
send out 9 verschicken, versenden, ausliefern
senior 114 vorgesetzt, leitend
sense of humour 115 Sinn für Humor
sense 59 Sinn
sensible 50 vernünftig
sensing coil 42 Sensorspule
sensor 112 Sensor, (Mess-)Fühler
sensory check 67 Sicht-, Geruchsprüfung
separate 30 getrennt
sequence 112 Reihenfolge
series 85 Reihe
in series 64p in Reihe (geschaltet)
serious 45 schlimm, schwer, ernst
serve 99 bedienen
serve as 23 dienen als
service 10 warten, pflegen
service entrance 26 Wartungs-, Inspektionszugang, -anschluss
service technician 94 Wartungstechniker/in
servicing instruction 64 (Betriebs-)Anleitung
set 71 (ein)stellen
a set of 17 eine Reihe von
be set for 109 eingestellt sein auf
set off 8 auslösen, hervorrufen
set up 8 eröffnen, einrichten
setting 111 Einstellung
settle 125 sich beruhigen
several 25 einige, mehrere
severe 45 schwer, stark
shade 59 beschatten
shaft 93 Achse, Welle
shake hands 8 sich die Hand geben
share 46 (sich) teilen, gemeinsam (be)nutzen
shave 125 (sich) rasieren
shaver socket 31 Steckdose für Rasierer

sheet 85 Blatt (Papier)
shift 15 Schicht
shining example 91 leuchtendes Beispiel
shiny 125 glänzend
(get a) shock 27, 34 (einen) Schock, Schlag (bekommen)
short circuit 70t Kurzschluss
shorten 74 (ver)kürzen
shortly 121 in Kürze
show around 14 herumführen
shower outlet 44 Duschanschluss
shrub 65 Busch, Strauch
shutdown 64 Abschaltung
shuttle train 110t Zubringerzug
on the other side of 14 gegenüber
side-cutting pliers 20 Seitenschneider
at first sight 59 auf den ersten Blick
sign 17 (Verkehrs-, Hinweis-)Schild
significant 65 bedeutend, spürbar
silicon 57 Silizium, Silikon
similar 18 ähnlich
similarity 103t Ähnlichkeit
simulate 75 simulieren
simultaneous(ly) 53 gleichzeitig
single socket 30t einfache Steckdose
single 30t einfach
sink unit 30t Spüle, Spülbecken
sinusoidal 78 sinusförmig
sir 21t mein Herr
site 8 (Bau-)Stelle
on site 107 auf der Baustelle, vor Ort
sketch 37 Skizze
skill 15 Fertigkeit, Fähigkeit
sleep mode 64 Bereitschaft, Ruhezustand
slide 53 schieben
slight 45 leicht
slip out of 95 herausrutschen
slot 53 Schlitz
slow down 46 (sich) verlangsamen
smooth 82 glätten
smooth(ly) 23 ruhig, rund
smoothing capacitor 83 Sieb-, Glättungskondensator
smoothing circuit 71 Glättungsschaltung
so far 124p bis jetzt
SOA (Safe Operation Area) control 85 Regelung des sicheren Arbeitsbereiches
so-called 101 so genannt
soccer 119 Fußball
sock it to s.o. 72 es jdm zeigen
socket 9 Steckdose
solar energy 10 Solarenergie
solar technology 10 Solartechnik
solar thermal collector 56 Sonnen(wärme)kollektor
solder 21 löten
solder in 71 anlöten
solder joint 70 Lötstelle
soldered 70 gelötet
soldering iron 20 Lötkolben
solenoid 41 Solenoid, Magnetspule
solid 15 fest, solide
solution 10 Lösung
solve 10 lösen
sophisticated 95 hoch entwickelt, kompliziert
sort out 109 klären, lösen
sound 8 anhören, klingen
source 40 Quelle
space 8 Platz, Fläche
spanner 20 Schraubenschlüssel
spare 23 Ersatz-
spark 8 Funke
special 74 Spezial(anfertigung)
specialize in 6 sich spezialisieren auf
specialist 58 Spezialist/in
specially designed 111 speziell entwickelt
specification 90 Angabe
specify 64 aufführen, angeben
speed 88 Geschwindigkeit
spin 87 schleudern, (schnell) drehen
split-ring commutator 93 Kommutator, Stromwender (mit Stromwenderlamellen)
spokesman 95 Sprecher
spokesperson 115 Sprecher/in
spot lamp 31 Spot
spot 30 herausfinden, bemerken
spray plate 44 Düsenkopf
spreadsheet 119 Tabellenprogramm, -kalkulation
square root 77 Quadratwurzel
square 69 Quadrat, Kästchen
stabilizer 82 Stabilisator
staff 12 Angestellte, Personal
stage 79 Phase, Stadium
stand-alone (computer) 46 Einzel(computer)
stand for 59 stehen für
standardization 96 Standardisierung
standardized 91 standardisiert, genormt
standby 67 Bereitschaft, Ruhezustand
star 49 Stern
stare into 115 anstarren
start button 23 Startknopf
statement 6 Aussage
state-of-the-art 10 auf dem neuesten Stand
stator 92 Stator, Ständer, feststehender Teil
status 103t Status, (soziale) Stellung
steady (-ily) 57 stetig
step by step 82 Schritt für Schritt
step down 35 reduzieren, heruntertransformieren
step through 110t durchgehen
step up 35 hochtransformieren
stereotype 103t Klischee(vorstellung)
stock 14 Vorrat, Lager(bestand)
in stock 21t vorrätig, auf Lager
stopcock 44 Absperrhahn
stopover 108 Zwischenlandung
storage 8 Lager(haltung)
store 9 lagern, aufbewahren

straight 41 gerade
straight away 22 sofort, gleich, auf der Stelle
strange 18 ungewöhnlich
strategy 103t Strategie
street lighting electrician 16 Elektriker/in für öffentliche Beleuchtung
strength 45 Kraft, Stärke
stressful 12 (sehr) anstrengend, stressig
strike 57 aufschlagen
strip 21 abisolieren
strip out 17 demontieren, herausnehmen, entfernen
strong-smelling 125 stark riechend
structure 97 Struktur
stuck 89 fest-, eingeklemmt
stupid(ly) 88 dumm, blöde
style 27 Stil
subcontractor 107 Subunternehmer
submit 15 einreichen
substation 35 Umspannwerk
subtract 61 abziehen, subtrahieren
successful 103t erfolgreich
suck off 71 absaugen
sudden 40 plötzlich
sufficient 64 ausreichend
suggest 27 vorschlagen
suggestion 50 Vorschlag
suitable (-ly) 25 geeignet, passend
suitable (for) 44 geeignet (für), passend
suitably rated 44 mit entsprechenden Nenndaten
sum 60p Summe, Betrag
summary 118 Zusammenfassung
sun 10 Sonne
sunlight 57 Sonnenlicht
supervise 109 beaufsichtigen
supervision 107 Aufsicht, Überwachung
supplied 44 mitgeliefert
supplier 52 Anbieter/in, Lieferant/in
supply 14 (Waren-)Angebot
supply with 51 (be)liefern, zur Verfügung stellen
support 19 unterstützen
surface 57 (Ober-)Fläche
surge 40 (plötzlicher) Anstieg
surround 53 umgeben
swap 51 (ver)tauschen
swear 95 schwören
switch 8 Schalter
switch off 29 ab-, ausschalten
switch on 43 einschalten
switchboard 117 Schalttafel
switched earthed socket outlet 31 schaltbare geerdete Steckdose
switched lamp 31 Leuchte mit Schalter

T

tag 29 kennzeichnen, etikettieren
take back 67 mitnehmen
take care 74 pass auf dich auf!
take for granted 97 als selbstverständlich betrachten/ansehen
take into account 57 in Betracht ziehen, berücksichtigen
take into consideration 102 berücksichtigen
take on 124 einstellen
take over 100 übernehmen
take part in 25 teilnehmen an, mitmachen bei
take place 97 stattfinden
take sth apart 72 etw auseinandernehmen
take turns 69 (sich) abwechseln
takeaway 65 zum Mitnehmen
taken 110t besetzt
talk about 103 sprechen über
talk sb through sth 59 jdm etw erläutern
talk to 30 sprechen mit
tank 37 Wasserspeicher
tankless 37 ohne Speicher
tape recorder 124p Tonbandgerät
target 123t Ziel
task 17 Aufgabe, Arbeit
technical 11 technisch
technical drawing 18 technische Zeichnung
technological 57 technologisch
telecommunications 98 Fernmeldewesen
temperature indicator 112 Temperaturanzeige
temperature 41 Temperatur
tense 63 Zeit(form)
term 15 Bedingung
terminal 29 Ausgang, Anschluss
terrible 84 schrecklich, furchtbar
test button 42 Prüf-, Testknopf
test light 76 Prüfleuchte
textile mill 106 Textilfabrik
theoretical 121 theoretisch
thermal circuit-breaker 41 thermischer Schutzschalter
thermal collector 57 Wärmekollektor
thermal-magnetic circuit-breaker 41 thermisch-magnetischer Schutzschalter
thermostat 106 Thermostat
thorough(ly) 125 gründlich
thousand 60 tausend
throughout 101 überall (in)
throw away 71 wegwerfen
thumb 91 Daumen
tidal 56 Gezeiten-
tidy (up) 65 aufräumen
tie 91 befestigen, binden
tie one's hair back 34 das Haar zusammenbinden
tight(ly) 53 fest
tighten 21 fest-, anziehen, festschrauben
tights 125 Strumpfhosen
time off (work) 25 (arbeits)freie Zeit, arbeitsfrei
on time 103t pünktlich
time-consuming 46 zeitraubend
timer 112 Zeitregler
timer control 87 Zeitschalter
timing 112 Zeitplanung

tin 65 Dose, Büchse
tingle 45 Prickeln, Kribbeln
to be on the safe side 37 um sicherzugehen
to date 107 bisher
tone 75 Ton
tool 20 Werkzeug, Instrument, Gerät
topic 98 Thema
topology 49 Topologie, Architektur
torque 93 Drehmoment, -kraft
get in touch with 25 (sich) in Verbindung setzen mit
touch-oriented 103t kontaktorientiert
towards 55 in Richtung auf
towel 95 Handtuch
trace 68 finden, zurückverfolgen
track 72 Stück, Song
trade fair 103 Handelsmesse
trade 56 Fach
trading 97 Geschäfts-, Handels-
traffic 17 (Straßen-)Verkehr
traffic light 17 Ampel
trainee 6 Auszubildende/r, Praktikant/in
traineeship 120 Ausbildung
training 22 Ausbildung
training allowance 25 Ausbildungsbeihilfe
training course 15 Lehrgang
transfer 111 Übermittlung
transfer 57 weiterleiten
transform 93 um-, verwandeln
transformer 35 Transformator
transistor 66 Transistor
translation 28 Übersetzung
transmission 91 Triebwerk, Getriebe
transmission shaft 91 Antriebs-, Getriebewelle
transmitter 75 Sender
transport 35 transportieren
trap 57 einfangen
travel around 17 herumfahren
tray 95 Schlitten
treat 95 behandeln
trendy 125 modisch, schick
triangle 101 Dreieck
trigger 42 auslösen
trip 34 betätigen, auslösen
triple 30t dreifach, Dreifach-
get into trouble 39 Probleme bekommen
troubleshoot 17 Störung suchen
tub 88 (Wasch-)Trommel
tube (AE = valve BE) 75 Röhre
tube amplifier 75 Röhrenverstärker
tumble dryer 65 Wäschetrockner
in turn 94 der Reihe nach
turn off 65 ausschalten
turn on 34 anschalten
turn the floor over to 115 das Wort übergeben an
two-way switch 29 Wechselschalter
typical(ly) 55 normal(erweise)

U

ugly 37 hässlich
unacceptable 41 unannehmbar, inakzeptabel
unattended 110 unbeaufsichtigt
unauthorized 55 unbefugt, unberechtigt
unavailable 55 nicht verfügbar
uncomfortable 115 unwohl
uncontrolled 40 unkontrolliert
uncoordinated 45 unkoordiniert
under guidance 111 unter Leitung
under 11 unter
underground 35 unterirdisch
undersink 38 Unterbau-
ungrounded 101 nicht geerdet
unique 75 einzigartig, einmalig
unless 64 wenn nicht, es sei denn
unlike poles 93 ungleiche Pole
unnecessary 55 unnötig
unpleasant 45 unangenehm
unplug 74 den Stecker herausziehen
unplugged 72 unplugged, ohne Verstärker
unqualified 14 nicht qualifiziert
unscrew 27 heraus-, abschrauben
unsolder 71 entlöten
unusual 88 ungewöhnlich
upload 117 senden
up-to-date 51 aktuell, auf dem neuesten Stand
user manual 95 Benutzerhandbuch
utility 57 Versorgungsbetrieb

V

V/t graph 78 Spannungs-Zeit-Diagramm
vacancy 121 offene Stelle
vacuum cleaner 93 Staubsauger
vacuum tube technology 75 Röhrentechnologie
valuable 46 wertvoll
valve 66 Ventil
van 10 (Werkzeug-)Wagen
variety 17 Vielfalt, Abwechslung
various 124p verschiedene
vary 75 sich unterscheiden, verändern, variieren
vehicle 13 Fahrzeug
ventilation 53 Lüftung, Belüftung
ventilation system 17 Belüftungssystem
ventricular fibrillation 45 Herzflimmern, Kammerflattern
venue 107 (Tagungs-)Ort
vernier calliper 20 Messschieber
versatile 95 vielseitig
version 32 Version, Fassung
vertical (Y) axis 78 vertikale (Y-)Achse
via 35 über, durch
virus 55 Virus
visible 57 sichtbar
vital 97 wesentlich, unerlässlich, unbedingt notwendig
vocabulary 26 Vokabular, Wortschatz
vocational 119 Berufs-

vocational college (GB), vocational school (US) 120 Berufsschule, -kolleg, gewerbliche Schule
voltage 15 Spannung
voltage divider 85 Spannungsteiler
voltmeter 76 Spannungsmessgerät, -messer, Voltmeter
volume 95 Volumen, Menge, Umfang
voluntary 97 freiwillig
vulgar 105 vulgär

W

wage 15 Lohn, Gehalt
walk-in wardrobe 31 begehbarer Kleiderschrank
warehouse 8 Lager(halle, -haus)
warehouse clerk 11 Lagerarbeiter
warehouseman 11 Lagerist/in
warm up 6 Aufwärmphase
warning 34 Warnung, Warnhinweis
wash motor 88 Waschmotor
waste 59 verschwenden
watch out for 115 aufpassen auf
water heater 36 Wasserkocher
water pump pliers 20 Wasserpumpenzange
wattage 61 Wattleistung
wave 75 Welle
waveform 78 Wellenform
wavelength 75 Wellenlänge
weak 98 schwach
weakness 123t Schwäche
wear 53 Abnutzung
weaving machine 106 Webmaschine
web page 19 Internetseite
webmaster 120 Webmaster
on a weekly basis 107 wöchentlich
welding 106 Schweißen, Schweiß-
whereas 40 wohingegen, während
whole 10 ganz
willing 95 gewillt
willing(ly) 98 bereit(willig)
wind 42 (sich) winden, biegen
wind turbine 63 Windturbine, -rad
window seat 110t Fensterplatz
wire 6 Leitungen verlegen
wire strippers 20 Abisolierzange
wireless 47 drahtlos, Funk-
wirer 117 Kabelleger/in
wiring plan 17 Verdrahtungs-, Leitungsplan
wiring 11 Verkabelung
be with sb in a moment 125 gleich bei jdm sein
without delay 10 unverzüglich
WLAN (Wireless Local Area Network) 47 lokales drahtloses Funknetz
word processing 119 Textverarbeitung
work experience 118 Berufserfahrung
work out 77p ausrechnen
work placement 124p Praktikum
work site 17 Baustelle
work with 15 arbeiten mit
workable 97 anwendbar
workbench 20 Werkbank
working conditions 17 Arbeitsbedingungen
working week 17 Arbeitswoche
workman 34 Handwerker
workmate 110t (Arbeits-)Kollege
workplace 16 Arbeitsplatz
worksheet 61 Arbeitsbogen
workshop 17 Werkstatt, Werkshalle
workstation 49 Arbeitsplatz (mit Computer)
worldwide 75 weltweit
be worried about 47 sich Sorgen machen wegen
be worth 23 wert sein, lohnen
be worthwhile 87 sich lohnen
write up 82 aufschreiben
writing 80 Schrift
written 108 schriftlich
be wrong about 19 Unrecht haben mit/bei

Y

Yours sincerely 7 Mit freundlichen Grüßen

Z

zigzag 24 zickzackförmig

BASIC WORD LIST

Diese Liste enthält ca. 700 Grundwörter, die in *Electricity Matters* als bekannt vorausgesetzt werden. Nicht aufgeführt, jedoch vorausgesetzt sind internationale Wörter (*taxi*, *email* usw.) und einige sehr elementare Wörter wie Pronomen, Zahlen und Tage.

a bit ein bisschen
a few ein paar
a lot viel, sehr
about über, etwa
above über, oben
abroad im/ins Ausland
accommodation Unterkunft
across (quer) über
action Handlung
active aktiv, tätig, aktiviert
activity Tätigkeit
actually tatsächlich
add zusammenzählen, hinzufügen
address Adresse; adressieren
advantage Vorteil
advertisement Anzeige
after nach
afternoon Nachmittag
again wieder
against gegen
age Zeitalter, Alter
ago vor
air Luft
airport Flughafen
alcoholic alkoholisch
all alle(s)
all right in Ordnung
almost fast, beinahe
along entlang
already schon, bereits
alright in Ordnung
also auch, außerdem
although obwohl
always immer
American amerikanisch; Amerikaner/in
and so on und so weiter
another noch eine
answer Antwort, Lösung; (be)antworten
any irgendetwas, -welche, jede
anyone jemand, jede/r
anything etwas, alles
anytime jederzeit
anyway jedenfalls, sowieso
anywhere irgendwo(hin)
apartment Wohnung
area Gebiet, Bereich
around herum
arrive ankommen
article Artikel, Gegenstand
as wie, als, da
as soon as possible so bald wie möglich
as well auch
ask (for/about) fragen (nach), bitten (um), sich erkundigen (nach)
at home zu Hause
at least mindestens
at the end am Ende, schließlich
atmosphere Atmosphäre
authority Autorität
automatic automatisch
automobile Auto(mobil)
away weg, entfernt
back zurück
bad schlecht, schlimm
bag Tasche
bath Bad
bathroom Bad(ezimmer)
be able to können
be afraid sich fürchten, Angst haben
be called heißen
because weil
become werden
bed Bett
bedroom Schlafzimmer
beer Bier
before vor(her)
begin anfangen, beginnen
beginning Anfang, Beginn
behave benehmen
behind hinter, hinten
below unter, unten
best beste/r/s, am besten
better besser
between zwischen
big groß
birthday Geburtstag
blouse Bluse
body Körper
book Buch; buchen, bestellen
bored gelangweilt
both beide
bottle Flasche
box Kasten, Kästchen, Schachtel
bring bringen, holen
British britisch
build (auf)bauen
building Gebäude
burn brennen
bus Bus
but aber, sondern
button Taste, Knopf
buy kaufen
by bei, durch, neben
call Anruf; (an)rufen
can dürfen, können
canteen Kantine
car park Parkplatz
car Auto
careful vorsichtig, sorgfältig
central zentral
centre Zentrum
century Jahrhundert
chance Chance, Gelegenheit
change Wechsel, (Ver-)Änderung; (aus)wechseln, (sich) ändern
cheap billig, günstig
check (über)prüfen
children Kinder
Chinese chinesisch
choose (aus)wählen
church Kirche
class Klasse
classroom Klasse(nzimmer)
clean reinigen, säubern; sauber
clear klar, deutlich
close schließen
clothes Kleidung, Kleider
cloudy bewölkt
coffee Kaffee
cold kalt
colleague Kollege, Kollegin
come kommen, geliefert werden
community Gemeinde
computer game Computerspiel
conservative konservativ
constant konstant
conversation Unterhaltung
cool kühl; kühlen
copy Kopie; kopieren
correct richtig, genau; korrigieren
cost Kosten; kosten
country Land, Staat
cultural kulturell
culture Kultur
cup Tasse
cupboard (Geschirr-)Schrank
cut schneiden
daily täglich
daily calendar (Termin-)Kalender
danger Gefahr
dangerous gefährlich
dark Dunkelheit; dunkel
date Datum, Termin
date of birth Geburtsdatum
day Tag

dead tot
dear liebe/r
death Tod
decision Entscheidung
describe beschreiben
dialogue Dialog
dictionary Wörterbuch
different anders, verschieden
difficult schwer, schwierig
digital digital
dinner (Abend-)Essen
direct direkt, gerade
disadvantage Nachteil, Schaden
dish Schüssel, Gericht
do tun, machen
doctor Arzt, Ärztin
document Dokument, Papier
door Tür
down unten
draw ziehen, zeichnen
drink Getränk; trinken
drive (Auto) fahren
driver Fahrer/in
dry trocken
during während
Dutch holländisch
each jede/r/s
early früh
easy leicht, einfach
eat essen
education Erziehung, Bildung
either ... or entweder ... oder
electricity Elektrizität, Strom
end Ende, Schluss; (be)enden
energy Energie
English englisch
enough ausreichend, genug
enter eintreten, eingeben
European europäisch; Europäer/in
even sogar (noch)
even if selbst wenn
evening Abend
every jede/r/s
everybody jede/r
everyday jeden Tag, alltäglich
everyone jede/r/s, alle
everything alles
example Beispiel
exciting aufregend, spannend
exercise Übung
exist existieren, bestehen
exit Ausgang
expensive teuer
expert Fachmann/frau
explain erklären
eye Auge
fall fallen
false falsch
family Familie
far weit (entfernt)
fast schnell
father Vater
fault Fehler, Mangel, Defekt
favourite Liebling(s-)
feel (sich) fühlen
field Feld, Bereich
fill füllen
fill in eingießen
find finden, suchen
find out herausfinden
fine gut, schön, fein
finger Finger
finish Ende, (be)enden
firm Firma, Unternehmen
first of all zu(aller)erst
firstly erst; zuerst
flight Flug
floor Etage, Stock(werk), (Fuß-)Boden
fly fliegen
follow (be)folgen
food Essen, Nahrung
for für
for example zum Beispiel
foreign ausländisch
forget vergessen
form Form, Formular
formal formell, förmlich
free kostenlos, frei
French französisch
fresh frisch
friend Freund/in
from von
garage
garden Garten
gas Gas, Benzin
gentlemen Herren
German deutsch; Deutsche/r
get holen, bekommen, werden
give geben
glass Glas
glasses Brille
go back zurückgehen
good gut
good afternoon guten Tag
good evening guten Abend
good luck viel Glück
good morning guten Morgen
grammar Grammatik
grass Gras
great groß(artig)
greet (be)grüßen
group Gruppe
guest Gast
guitar Gitarre
hair Haar(e)
half Hälfte; halb
hang (auf)hängen
happy glücklich, froh, zufrieden
hard schwer
hate hassen, nicht mögen
have haben
have a look at (sich) ansehen
head Kopf
heading Überschrift
hear hören
heavy schwer
help Hilfe; helfen
helpful hilfreich, nützlich
here hier
high hoch; äußerst
holiday Ferien, Urlaub, Feiertag
homework Hausarbeit, -aufgabe(n)
honest ehrlich, anständig
hope hoffen; Hoffnung
hospital Krankenhaus
hot heiß, warm
hour Stunde
hourly stündlich
house Haus
how wie
hungry hungrig
idea Idee, Gedanke
if wenn, falls, ob
illustration Abbildung
impatient ungeduldig
important wichtig
in in
in fact tatsächlich
in particular besonders
individual einzeln, individuell
industry Industrie
infinitive Infinitiv
info, information Auskunft, Information(en)
inside innerhalb, drinnen
instead (of) stattdessen, anstatt
interest Interesse, Zins(en); interessieren
interesting interessant
international international
Italian italienisch; Italiener/in
Japanese japanisch
journalist Journalist/in
just einfach, nur, genau
keep (be)halten
key Schlüssel, Taste
kill töten
kind of Art von
kitchen Küche
knowledge Wissen, Kenntnis(se)
lady Dame
land Land
language Sprache
large groß, umfangreich
last letzte/r/s, zuletzt
late spät
lead führen
leader Führer/in
learn lernen, erfahren
left linke/r/s, links; übrig
less weniger, abzüglich
lesson (Unterrichts-)Stunde

letter Brief, Buchstabe
level Niveau
life Leben
light Licht, Leuchte; hell, leicht
like (ähnlich) wie; mögen, gern tun
line Linie, Leitung, Zeile
list Liste; auflisten, notieren
listen to zuhören
listener (Zu-)Hörer/in
literature Literatur
little klein, wenig
live wohnen, leben
living room Wohnzimmer
long lang
look Blick; (aus)sehen, blicken
look for suchen nach
lose verlieren
lots of viel, viele
loud laut
love Liebe; lieben, sehr gern mögen
low niedrig
lunch Mittagessen
machine Maschine
magazine Zeitschrift
main hauptsächlich, wichtigste
mainly hauptsächlich
make machen
man Mann, Mensch
management Verwaltung, Management
manager Manager/in
many viele
map Karte
market Markt
match zuordnen, (zusammen) passen; Spiel
matter Sache
maximum Maximum; maximal
maybe vielleicht
mean bedeuten, meinen, heißen
medium-sized mittelgroß
meet (zusammen)treffen, begegnen
meeting Konferenz, Besprechung
metal Metall
middle Mitte
midnight Mitternacht
mineral water Mineralwasser
minimal minimal
minute Minute
miss verpassen, vermissen
mistake Fehler, Irrtum
model Modell, Model
moment Moment, Augenblick
money Geld
month Monat
monthly monatlich
more mehr
more than mehr als
morning Morgen
most meist
move (sich) bewegen, umziehen
much viel
music Musik
musical instrument Musikinstrument
musician Musiker/in
Muslim Muslim/in
must müssen
name Name; (be)nennen
national national, staatlich
nationality Staatsangehörigkeit
nature Natur
near nahe
nearly beinahe, fast
need Bedarf; brauchen, benötigen
negative negativ, verneinend
never nie(mals)
new neu
newspaper Zeitung
next nächst; danach
nice schön, nett
night Nacht
nobody niemand
normal normal
north Norden
north-west Nordwesten
note Notiz; beachten, notieren
nothing nichts
notice Notiz
now nun, jetzt
number Nummer, Zahl
o'clock (...) Uhr
object Gegenstand
of course natürlich, selbstverständlich
offer Angebot; (an)bieten
office Büro, Amt
officer Beamte/r
official amtlich
often oft, häufig
okay okay
old alt
on auf, an, am
once einmal, einst, sobald
only nur, einzig
open öffnen, beginnen; offen
operation Operation, Handlung
opinion Meinung, Einschätzung
opposite Gegenteil; gegensätzlich
or oder
organization Organisation
other andere/r/s
out aus
outside außer(halb)
over (vor)über
own eigene; besitzen
page Seite
pair Paar
parent Eltern(teil)
part Teil, Rolle
party Party, Partei
passive passiv; Passiv
passport Pass
past vorbei, nach; Vergangenheit
pay (be)zahlen
people Personen
perfume Parfüm
person Mensch, Person
phone Telefon, anrufen
phone call (Telefon-)Anruf
photo(graph) Foto(grafie)
picture Bild
piece Stück, Teil
place Stelle, Platz; setzen, stellen
plan Plan; planen
plane Flugzeug
play spielen
please bitte
pleased zufrieden
pocket (Hosen-)Tasche
point Punkt
position Stellung
positive positiv, bejahend
possibility Möglichkeit
possible möglich, denkbar
pot Topf, Becher
practice Übung, Praxis
present Geschenk
pretty hübsch
price (Kauf-)Preis
primary school Grundschule
print drucken
probable wahrscheinlich
probably wahrscheinlich
produce produzieren, herstellen
product Produkt, Erzeugnis
production Produktion, Herstellung
professional professionell, beruflich
programme Programm
project Projekt, Vorhaben
protect (be)schützen
push schieben
put setzen, stellen, legen
quality Qualität
question Frage
quick schnell, rasch
quite ziemlich, ganz
railway (Eisen-)Bahn
railway station Bahnhof
rain Regen; regnen
reaction Reaktion
read lesen
ready fertig, bereit
real(ly) echt, wirklich
reason Vernunft, Grund
recently neulich, kürzlich
regular regelmäßig, regulär
remember sich erinnern, daran denken
reply Antwort, antworten
rest Rest
restaurant Restaurant
result Resultat, Ergebnis; folgen

return zurückkehren, -gehen
revolution Revolution, Umdrehung
right rechts, rechte/r/s, richtig;
ring Ring; klingeln, läuten
river Fluss
rock Felsen, Stein
role Rolle
room Zimmer, Raum
run laufen; leiten
safe sicher
safety Sicherheit
same gleiche/r/s, der-, die-, dasselbe
say sagen
school Schule
Scottish schottisch
search Suche; (durch)suchen
secretary Sekretär/in
see sehen
seem (er)scheinen
sell (sich) verkaufen
send senden, schicken
sentence Satz
sharp scharf, spitz
shirt Hemd
shoe Schuh
shop Laden, Geschäft; einkaufen
short kurz, klein, knapp
should solle/n, sollte/n
shoulder Schulter
show zeigen
shower Dusche
side Seite, Rand
sign (An-)Zeichen
simple einfach
since da, weil, seit
sister Schwester
sit sitzen, sich hinsetzen
sit down sich (hin)setzen
situation Situation, Lage
size Größe
skin Haut
slow langsam, verlangsamen
small klein
smell Geruch; riechen
smile Lächeln; lächeln
so also, damit, deshalb, so (dass)
sock Socke; schlagen
some einige, etwas
someone jemand
someone else jemand anders
something etwas
sometimes manchmal
somewhere irgendwo(hin)
soon bald
sorry traurig; Verzeihung
sort of Art von
soup Suppe
south Süden
Southern südlich
speak sprechen, reden
speaker Sprecher/in, Redner/in
spend ausgeben (Geld), verbringen (Zeit)
sport Sport
standard Norm, Standard, Maßstab; normal, üblich
start Beginn; anfangen, starten, beginnen
state Zustand; feststellen
stay Aufenthalt; bleiben
step Schritt
still still, trotzdem, (immer) noch
stop Halt; (an) halten, aufhören (mit)
street Straße
strong stark, kräftig
student Student/in, Lernende/r
study lernen, studieren, untersuchen, betrachten
subject (Schul-)Fach, Thema
such so, solch
such as wie zum Beispiel
suddenly plötzlich
summer Sommer
sure freilich, sicher(lich)
surprise Überraschung; überraschen
sweater Pullover
table Tisch, Tabelle
take nehmen, bringen, dauern
talk (about) Gespräch, Unterhaltung, Vortrag; sprechen, reden (über)
teach unterrichten, lehren
teacher Lehrer/in
team Mannschaft
technology Technik, Technologie
teeth Zähne
telephone Telefon
tell sagen, erzählen
test Test, Prüfung; untersuchen, prüfen
text Text
than als
thank danken
thanks Dank; danke
that das, dass
then dann
there da, dort(hin)
therefore deshalb
these diese
thing Sache, Ding, Gegenstand
think denken, meinen, finden, glauben
thirsty durstig
this dies, diese/r/s
those jene
through durch
thus so
time Zeit, Mal
title Titel
today heute
together zusammen
tomorrow morgen
too zu, auch
top Spitze, Gipfel; Spitzen-
total (End-)Summe; völlig, (ins)gesamt
touch anfassen, berühren
tour Tour; bereisen
town Stadt
traditional traditionell
train Zug
translate übersetzen, -tragen
travel Reisen; reisen, fahren
traveller Reisende/r
tree Baum
true richtig, wahr
try versuchen, probieren
Turkish türkisch
turn Drehung, Wendung; (sich) drehen
type erfassen, tippen
typical typisch
ultraviolet ultraviolett
umbrella Schirm
under unter
understand verstehen, begreifen
uniform Uniform
unimportant unwichtig
unit Unit
universal universell, allgemein
unlike anders als, im Gegensatz zu
until bis
up auf
up to bis zu
use benutzen, verwenden
useful nützlich
user Anwender/in, Benutzer/in
usual(ly) normal(erweise), gewöhnlich
vegetable Gemüse
very sehr
visit Besuch; besuchen, besichtigen
visitor Besucher/in, Gast
wait (for) warten (auf)
walk Spaziergang; (zu Fuß) gehen
wall Wand, Mauer
want wollen
warm warm
warn warnen
wash waschen, spülen
watch (Armband-)Uhr
water Wasser
way Weg, Methode, Art (und Weise)
wear tragen, anhaben
weather Wetter
website Website
week Woche
weekend Wochenende
welcome willkommen
well gesund, gut, also
Western westlich
wet nass
what was, welche/r/s

when wenn, als, wann
where wo(hin)
whether ob
which welche/r/s
while während
who wer, der/die/das, welche(r/s)
why warum
wide breit, weit
wife (Ehe-)Frau
window Fenster
wish Wunsch; wünschen
with mit, bei
within innerhalb (von), in
without ohne
woman Frau
wonder Wunder; sich fragen
word Wort
work Arbeit, Werk; funktionieren, arbeiten
world Welt
worry sich Sorgen machen
worse schlechter, schlimmer
worst schlechteste, schlimmste
write (down) (auf)schreiben
wrong falsch
year Jahr
yesterday gestern
yet noch
young jung

IRREGULAR VERB LIST

be	was/were	been	*sein*
become	became	become	*werden*
begin	began	begun	*anfangen*
break	broke	broken	*brechen*
bring	brought	brought	*bringen*
build	built	built	*bauen*
burn	burnt/burned	burnt/burned	*brennen*
buy	bought	bought	*kaufen*
come	came	come	*kommen*
cost	cost	cost	*kosten*
do	did	done	*tun, machen, erledigen*
draw	drew	drawn	*zeichnen*
drink	drank	drunk	*trinken*
drive	drove	driven	*fahren*
eat	ate	eaten	*essen*
feel	felt	felt	*sich fühlen*
find	found	found	*finden*
forget	forgot	forgotten	*vergessen*
get	got	got	*bekommen, erhalten*
go	went	gone	*gehen*
grow	grew	grown	*wachsen*
have	had	had	*haben*
hear	heard	heard	*hören*
hold	held	held	*halten*
keep	kept	kept	*behalten*
know	knew	known	*kennen, wissen*
lay	laid	laid	*legen*
learn	learned/learnt	learned/learnt	*lernen*
leave	left	left	*verlassen*
lend	lent	lent	*(aus)leihen*
let	let	let	*lassen*
lose	lost	lost	*verlieren*
make	made	made	*machen*
mean	meant	meant	*bedeuten*
meet	met	met	*sich treffen*
pay	paid	paid	*(be)zahlen*
put	put	put	*setzen, stellen, legen*
read	read	read	*lesen*
ring	rang	rung	*läuten, klingeln, anrufen*
rise	rose	risen	*steigen*
run	ran	run	*laufen*
say	said	said	*sagen*
see	saw	seen	*sehen*
sell	sold	sold	*verkaufen*
send	sent	sent	*senden, schicken*
show	showed	shown/showed	*zeigen*
sit	sat	sat	*sitzen*
speak	spoke	spoken	*sprechen*
spend	spent	spent	*ausgeben, verbringen (Zeit)*
stand	stood	stood	*stehen*
take	took	taken	*nehmen*
teach	taught	taught	*unterrichten, lehren*
tell	told	told	*erzählen, mitteilen, sagen*
think	thought	thought	*denken, meinen*
understand	understood	understood	*verstehen*
write	wrote	written	*schreiben*

ABBREVIATIONS

A	amp(ere)	*Ampere*
AE	American English	*amerikanisches Englisch*
a.m.	ante meridiem	*morgens*
AC	alternating current	*Wechselstrom*
BE	British English	*britisches Englisch*
C	centigrade, Celsius	*Celsius*
cm	centimentre	*Zentimeter*
CO_2	carbon dioxide	*Kohlensäure*
CV	curriculum vitae	*Lebenslauf*
dB	decibel	*Dezibel*
DC	direct current	*Gleichstrom*
DVD	digital video disc	*DVD*
e.g.	for example	*zum Beispiel*
Enc	enclosure	*Anlage*
etc	et cetera	*und so weiter*
F	farad	*Farad*
FM	frequency modulation	*Frequenzmodulation*
G	giga- (one billion)	*Giga-*
GB	Great Britain	*Großbritannien*
GFCI	ground fault circuit interrupter	*FI-Schalter*
H	henry	*Henry*
Hz	hertz	*Hertz*
in	inch	*Zoll*
k	kilo- (one thousand)	*Kilo-*
km	kilometre	*Kilometer*
kWh	kilowatt hour	*Kilowattstunde*
l	litre	*Liter*
LAN	local area network	*lokales Netz, LAN*
lb	pound (Gewicht)	*Pfund*
lbf	pounds force	*Pfundkraft*
LED	light emitting diode	*Leuchtdiode*
M	mega- (one million)	*Mega-*
m	metre	*Meter*
m	milli- (one thousandth)	*Milli-*
m^2	square metre	*Quadratmeter*
mm	millimetre	*Millimeter*
no.	number	*Nummer*
p.m.	post meridiem	*nachmittags, abends*
PCB	printed circuit board	*Leiterplatte*
plc	public limited company	*≈ Aktiengesellschaft (AG)*
PV	photovoltaic	*photovoltaisch, Photovoltaik*
R	resistance	*Widerstand*
RCD	residual current device	*FI-Schalter*
ref.	reference	*Betreff*
rms	root mean square	*Effektivwert*
rpm	revolutions per minute	*Umdrehungen pro Minute*
UHF	ultra-high frequency	*Ultrahochfrequenz*
US	the United States	*die Vereinigten Staaten*
V	volt	*Volt*
VAT	value added tax	*Mehrwertsteuer*
W	watt	*Watt*
WLAN	wireless local area network	*kabelloses lokales Netz, WLAN*

NUMBERS AND SYMBOLS

Cardinal numbers

0	zero/nought (BE)
1	one
2	two
3	three
4	four
5	five
6	six
7	seven
8	eight
9	nine
10	ten
11	eleven
12	twelve
13	thirteen
14	fourteen
15	fifteen
16	sixteen
17	seventeen
18	eighteen
19	nineteen
20	twenty
21	twenty-one
22	twenty-two
23	twenty-three
24	twenty-four
25	twenty-five
26	twenty-six
27	twenty-seven
28	twenty-eight
29	twenty-nine
30	thirty
40	forty
50	fifty
60	sixty
70	seventy
80	eighty
90	ninety
100	one hundred
101	one hundred and one
200	two hundred
210	two hundred and ten
1000	one thousand
2000	two thousand
10,000	ten thousand
10,532	ten thousand, five hundred and thirty-two
100,000	one hundred thousand
1,000,000	one million
1,000,000,000	one billion

Ordinal numbers

1st	first	**16th**	sixteenth
2nd	second	**17th**	seventeenth
3rd	third	**18th**	eighteenth
4th	fourth	**19th**	nineteenth
5th	fifth	**20th**	twentieth
6th	sixth	**21st**	twenty-first
7th	seventh	**22nd**	twenty-second
8th	eighth	**30th**	thirtieth
9th	ninth	**40th**	fortieth
10th	tenth	**50th**	fiftieth
11th	eleventh	**60th**	sixtieth
12th	twelfth	**70th**	seventieth
13th	thirteenth	**80th**	eightieth
14th	fourteenth	**90th**	ninetieth
15th	fifteenth	**100th**	one hundredth

Fractions

½	a/one half
⅓	a/one third
¼	a/one quarter
⅘	four fifths
⅜	three eighths
1/100	one hundredth
7½	seven and a half
6¾	six and three quarters

Decimals

0.1	zero/nought (BE) point one
2.5	two point five

Symbols

µ	micro- (one millionth)
Ω	ohm
%	per cent
°	degrees
+	plus
–	minus
±	plus or minus
×	multiplied by / times
*	multiplied by / times
÷	divided by
√	(the) square root (of)
=	is equal to / equals
≠	is not equal to / does not equal
≈	is approximately equal to
<	is less than
>	is more than

ACKNOWLEDGEMENTS

Fotos
RF-FOTOS M. Benford: S.16/1; S.17/2; S.32; S.44; S.66/6; S.67/2;
Corbis: S.6/1; S.10/a; S.10/f; S.17/1; S.17/3; S.17/4; S.102/2; Getty Images: S.110/WV;
H. Goebel: S.46; S.52/1,2,4,5,6; V. Maly: S.86;
Wolf-Rainer Windisch: S.28/1-8; S.29; S.39; S.41; S.52/3; S.66/1-5,7-10; S.76; S.77; S.78; S.81; S.82; S.83; S.92

Corbis: S.16/4/B.Rowan; S.57/W.Schroll; S.59/H.Mollenhauer; S.67/J.Sugar; S.102/1/P.Fisher; S.110/4/J.V.Hasselt; S.114/ R.Nowitz;
Fotofinder: S.6/2 VISUM/W. Steche; 6/3 Bilderbox/E.Wodicka; S.6/4 Introfoto/D.Ausserhofer; S.10/b Photothek/ T.Koehler; S.10/c VISUM/A.Buellesbach; S.10/d Keystone/V.Schulz; S.16/2 F1-Online/Widmann; S.16/3 Das Fotoarchiv/ M.Matzel; S.25 Keystone/J.Zick; S.106/1 Das Fotoarchiv/M.Vollmer; S.106 Das Fotoarchiv; S.110/1 Introphoto/ R.Steussloff; S.110/3 pictures News/M.Vedder; S.123 mediacolors;
Picture Alliance: S.10/e dpa/W.Baum;

Mit freundlicher Unterstützung folgender Unternehmen: S.20 Beck Elektrotechnik GmbH, Würzburg; S.22/König & Bauer AG, Würzburg; S.23/SKF GmbH, Schweinfurt; S.38/Clage GmbH, Lüneburg; S.87 Samsung; S.106/3 MRP Electronics; Bedford UK; S.106/4, S.111/Huppmann AG, Kitzingen;

Texte
S.53: Intel Corporation; S.55: MOREnet; S.65: BBC; S.70: http://de.wikipedia.org; S.85: http://www.alliedelec.com; S.91: http://www.applianceaid.com; S.93: www.bergen.org and http://en.wikipedia.org;
S.95: Washington Post, May 26, 2002; S.96 (Cartoon), S.97: Financial Times, October 14, 1994; S.112: http://en.wikipedia.org; S.125: Rowan Manahan, Where's My Oasis? Random House, 2004